RANDOM
HOUSE

LARGE
PRINT

THE CASE
FOR KETO

THE CASE
FOR KETO

Rethinking Weight Control and the Science

and Practice of Low-Carb/High-Fat Eating

Gary Taubes

RANDOM HOUSE
LARGE PRINT

Cover design: Jenny Carrow
Cover images: Germano Poli / EyeEm, Jenny Dettrick,
bluestocking, choness, JamieB, all Getty Images

Grateful acknowledgment is made to Oxford University
Press and Copyright Clearance Center for permission to
reprint previously published material from "The Heritage
of Corpulence" by E. B. Astwood, originally published
in Endocrinology (August 1962, Volume 71) by
Oxford University Press. Copyright © 1962 Oxford
University Press. Reprinted by permission of
Oxford University Press and Copyright Clearance Center.

The Library of Congress has established
a Cataloging-in-Publication record for this title.

ISBN: 978-0-593-29372-0

www.penguinrandomhouse.com/large-print-format-books

FIRST LARGE PRINT EDITION

Printed in the United States of America

10 9 8 7 6 5 4 3 2 1

This Large Print edition published in accord
with the standards of the N.A.V.H.

To Kitty and Larry

Contents

THE CASE
FOR KETO

Introduction

The conflict

I am not writing this book for the lean and healthy of the world, although I certainly believe they can benefit by reading it. I am writing it for those who fatten all too easily, who are drifting inexorably toward overweight, obesity, diabetes, and hypertension, or some combination of them, or who are already afflicted and are living at increased risk of heart disease, stroke, and, in fact, all chronic disease. And I'm writing it for their doctors.

This book is a work of journalism masquerading as a self-help book. It's about the ongoing conflict between the conventional thinking on the nature of a healthy diet and its failure to make us healthy, about the difference between how we have been taught to eat to prevent chronic disease and how we may have to eat to return ourselves to health. Should we be eating to reduce our risk of future disease, or should we be eating to achieve and maintain a healthy weight? Are these one and the same?

Since the 1950s the world of nutrition and chronic disease has been divided on these questions into two major factions. One is represented by the voices of authority, assuring us that they know what it means to eat healthy and that if we faithfully follow their advice, we will live longer and healthier lives. If we eat real food, perhaps mostly plants, and certainly in moderation, we will be maximizing our health. This advice goes along with the overwhelming consensus of opinion in the medical establishment that we get fat because we eat too much and exercise too little. Hence the means of prevention, treatment, or cure, whether provided by the pharmaceutical industry or by our own power of will, is to tame our appetites.

As I write this paragraph, the American Heart Association and the American College of Cardiology have just released their latest lifestyle guidelines. These health organizations recommend, as they have for decades now, that those who are fat or diabetic should restrict their calories, eat less (particularly less saturated fat), and perhaps take up regular exercise (or exercise more regularly) if they want to avoid premature death from heart disease. It all seems eminently reasonable—yet it clearly doesn't work, at least not on a population-wide basis. It likely hasn't worked for you if you're reading this book. This thinking, though, has been accepted as dogma for fifty years and is disseminated ubiquitously, even as the prevalence of obesity in the United States has increased by over 250 percent and diabetes by almost 700 percent (a number that I believe should frankly scare us all silly). So the question is, as it has always been, Is this thinking and advice simply wrong, or are we just not following it?

The other faction, the heretics, make their claims very often in the context of what the experts dismiss as fad diet books. These books offer up a very different proposition from the conventional thinking on healthy eating. While the authorities are telling us that if we eat as they propose, we will prevent or delay the eventual onset of chronic disease and live longer and healthier by doing so, these diet book doctors are claiming to be able to reverse chronic disease (including obesity) rather than prevent it. We should try their approach, these books imply, and see if it works: Does it help us achieve and maintain both health and a healthier weight? If it does, we can reasonably assume that it will lead as well to a longer and healthier life, heresy be damned.

The authors of these books claim to have confidence that their approach works, but we don't have to accept their words on faith. (Some of their advice is contradictory, so clearly it can't all work.) But if we can take their advice and get healthier and leaner by doing so, then each of us can decide if the consensus of medical opinion is right **for us** and perhaps at all.

The authors of these books almost invariably started their careers as practicing physicians, and many still are. Almost invariably, they say they struggled with their own excess weight but freed themselves from the conventional thinking long enough to delve into the research literature and seemingly solve the problem. They had what the journalist and best-selling author Malcolm Gladwell called in a 1998 **New Yorker** article, in precisely this context, a "conversion" experience. They found a way to eat that made it easy to achieve a healthy weight and then to maintain it. Then they tried it on their patients, and it worked (or so they claimed), and they wrote books about it, and the books often became best sellers.

These books are commonly based on a single fundamental assumption, sometimes implicit, sometimes explicit: We get fat not because we eat too much but because we eat carbohydrate-rich foods and drink carbohydrate-rich beverages. The culprits, specifically, are sugars, grains, and starchy vegetables. For those who fatten easily, these carbohydrates are the reason they do. One powerful implication of these diet books is that obesity is caused **not** by eating too much but by a hormonal imbalance in the body that eating these carbohydrate-rich foods triggers. It's a very different way of thinking about why we accumulate excess fat. It demands a very different approach to prevention and treatment.

Many if not most of the popular best-selling diets of the past forty years—Atkins, keto, paleo, South Beach, Dukan, Protein Power, Sugar Busters, Whole30, Wheat Belly, and Grain Brain—are or at least include variations on this simple theme: Specific carbohydrate-rich foods create a hormonal milieu in the human body that works to trap calories as fat rather than burn them for fuel. At the very simplest level, if we want to avoid being fat or return to being relatively lean, we have to avoid these foods. They are quite literally fattening.

Physicians now commonly refer to this way of eating as low-carbohydrate, high-fat (LCHF). At its extreme, it excludes virtually all carbohydrates other than those in green leafy vegetables and the tiny proportion in meat and is technically known as ketogenic, hence "keto" for short. I'll typically refer to it as LCHF/ketogenic eating to capture both concepts. The term has the great disadvantage of failing in any way to be catchy; it trips off no tongues. But it does have the advantage of being precise and inclusive in its meaning.

When I began my journalistic investigation into the convergence of diet, obesity, and chronic disease twenty years ago, perhaps a few dozen physicians in the world were openly prescribing LCHF/ketogenic eating to their patients. Today this philosophy and dietary prescription have been embraced by thousands of physicians, if not a few tens of thousands, more every day, for very simple reasons.* They are working on the front lines of the obesity and diabetes epidemics; they have a professional stake in seeing obesity and diabetes addressed correctly and reversed, if at all possible, by healthy dietary approaches. They do not have the luxury to treat their patients by offering them speculative, however well-accepted, hypotheses about the nature of a diet that might, according to statistical assessments, prevent heart attacks. Their patients are sick, and the goal of these physicians is to make them healthy.

Over the course of their careers, these doctors have seen their waiting rooms fill with patients who are ever more overweight, obese, and diabetic, as have doctors worldwide. Doctors told me in interviews that they went into medicine because they wanted to make people healthy and instead found themselves spending their days "managing disease," treating the symptoms of obesity and diabetes and the diseases associated with them ("comorbidities," in the medical jargon). They were becoming almost hopelessly discouraged. So they had a powerful incentive to shed their preconceptions about what **should** work, to renounce or at least

* In Canada alone, a Facebook group for women physicians on LCHF/ketogenic eating had over 3,800 members as of September 2019.

question the dietary dogma of their professional societies and their peers, and look for truly effective alternative solutions.

Almost invariably, these physicians had a personal stake as well. This is a critical point, and I will return to it: To accept the possibility that the conventional thinking on diet and weight is misconceived and so fails your patients, it helps to have experienced that failure yourself. Some of these physicians had been vegetarians for decades. Some had been vegans. Many are athletes, even ultra-endurance athletes. They prided themselves on eating "healthy" and yet found they had become fatter, diabetic, or prediabetic despite doing everything "right." They were telling their patients to eat low-fat diets, mostly plants, not too much (control their portion sizes), and to exercise. They were following that advice themselves—and it wasn't working.

Their rate of success in getting obese patients to lose meaningful amounts of weight with this diet and exercise prescription—as Deborah Gordon, a family medicine physician in Ashland, Oregon, described it to me—was "close to zero." So these doctors did what we would hope any thoughtful person would do, and certainly our physicians, in these circumstances: They kept their minds open and went searching for a better approach. When they read about LCHF/ketogenic eating—now easy to do on the Internet as well as in books—they opted to self-experiment. When they discovered that this way of eating worked for them, that it lived up to its promise, they had their conversion experience. Afterward they suggested it cautiously to their patients. When it worked for them—and they learned from experience what did and did not—they became passionate. These physicians became the founding members of a grassroots revolution that is working to change how we think about obesity and diabetes in America and around the world, and therefore how we prevent and treat them.

Take Susan Wolver, for instance, an air force flight surgeon turned internal medicine practitioner in Richmond, Virginia, and an associate professor at the Virginia Commonwealth University School of Medicine. Richmond happens to be among the fattest cities in the United States; a 2012 Gallup survey ranked it second in prevalence of obesity, behind only Memphis. As Wolver described it to me, all she did, seemingly day in and day out, was "take care of chronic diseases associated with obesity—hypertension, heart disease, diabetes." Wolver diligently advised her patients to eat healthy, eat less, and exercise, but her advice had little noticeable effect. By 2013, in her then twenty-three years in medicine, only two of her patients had lost significant weight following that advice, and one had very quickly regained it.

Throughout those years, Wolver assumed, as

doctors typically will, that her patients were not listening or were unwilling to make the necessary effort. "Then something happened," she said. "I got to be middle-aged. I was following the advice I had given to all my patients, but every time I stepped on a scale, it was clear my advice no longer worked for me. I had an epiphany: 'Maybe I'm wrong about my patients following my advice. Maybe my advice stinks.' I started a personal journey to see what works."

In 2012 Wolver began attending obesity and weight-loss sessions at medical conferences, hoping to learn anything plausible that she might try. At a day-long seminar hosted by the Obesity Society, she heard Eric Westman of Duke University Medical School present his clinical experience and research. Westman had done several of the earliest clinical trials comparing the kind of low-fat, portion-controlled, weight-loss diets advocated by the American Heart Association to the Atkins diet, an LCHF/ketogenic diet, restricted only in carbohydrates—in grains, in starchy vegetables like potatoes, and in sugars—and very rich in fat.

Westman reported that the Atkins diet allowed his patients to lose weight almost effortlessly and to become healthier in the process, just as Atkins had claimed. He said that it was confirmed not only by his patients' experiences but also by his own clinical trials and a growing list of others that had demonstrated that it was indeed a healthy way to eat.

"[Westman's] patients seemed a lot like mine," Wolver told me, with the difference that Westman's lost weight and kept it off while hers didn't. In May 2013 she drove two and a half hours south to Durham, North Carolina, and spent two days at Westman's clinic. She sat in on a day of follow-up visits and responded with "astonishment": "I'd never seen anything like it in my life: eighteen people that day. Seventeen had lost significant weight and kept it off. That was sixteen more than I had ever seen."

This is how unconventional or unorthodox practices spread through medicine. New drug therapies may become what physicians call "standard of care" when medical journals publish the latest clinical trial results, but the more mundane therapies (those, regrettably, that hold no promise of profiting the pharmaceutical or medical device industries or surgeons) spread initially by anecdote, observation, and clinical experience. One physician has a patient with a seemingly intractable medical condition and learns of another physician who may have a treatment that works. If it seems reasonably safe, she discusses the potential risks and benefits with her patient and gives it a try. If it works, she is likely to try it on others as well.

Two days after visiting Westman, Wolver was back in her Richmond clinic teaching her patients with obesity and diabetes to eat as Westman was teaching his. In the years since, she's given this dietary advice to over three thousand patients. Not only do her patients lose significant weight, just as Westman's do, but her diabetic patients get off their medications, often including insulin and blood pressure drugs. She said it's easier now than it was in her early years to convince her patients to buy in because resistance to the LCHF/keto approach has slowly eroded. And success breeds success. Every patient who loses weight and is taken off diabetes and blood pressure medications is an advertisement to friends, neighbors, coworkers, and family that they can do the same. Now Wolver gets referrals from local physicians, including cardiologists who would have feared until recently that the diet she recommends would increase risk of heart disease. Now they have compelling reason to believe it does the opposite. Over a third of her patients, Wolver said, are hospital employees, and they spread the word.

By prescribing to her patients what nutritional authorities would consider a fad diet, perhaps the most infamous of all fad diets, one rich in fat and saturated fat and restricted in all those carbohydrates that those authorities have insisted are heart-healthy diet foods, Wolver is making her patients healthy again. By prescribing this diet to her patients—an act that the Harvard nutritionist Jean Mayer equated in **The New York Times** in 1965 to "mass murder" and that the American Medical Association eight years later claimed to be based on "bizarre concepts of nutrition that should not be promoted to the public as if they were established scientific principles"—Wolver believes, as does Westman, that the benefits her patients are experiencing will translate to longer and healthier lives. So it spreads from physician to physician, and the unconventional slowly makes the transition to standard of care—because it works.

In the early 2000s, when I interviewed over six hundred clinicians, researchers, and public health authorities for my first book on nutrition science, **Good Calories, Bad Calories,** some of the most influential among them readily admitted to using the LCHF/ketogenic diet themselves. "It's a great way to lose weight," the renowned Stanford University endocrinologist Gerald Reaven said to me about the Atkins diet. "That's not the issue." But these physician-researchers would not prescribe it for their patients, thinking the risk of causing harm was too great. **That** was the issue. They would eat the fat-rich, ketogenic Atkins diet themselves until they lost their excess pounds; then they'd stop and eat "healthy." When they regained the weight, they would repeat the diet.*

---

* As I'll discuss, some authorities argued that the Atkins diet and those similar should never be recommended because they are too difficult to maintain. Jean-Pierre Flatt, a University of Massachusetts biochemist whose thermodynamic hypothesis of why we get fat

One significant difference between the physician researchers I interviewed in the early 2000s and those in clinical practice that I interviewed for this book—more than one hundred through the summer and fall of 2017 (plus a dozen or so dietitians and nurse practitioners, a few chiropractors, health coaches, and a dentist)—is that the latter believe these diets are inherently healthy, perhaps the healthiest way for many if not most of us to eat. In that sense, they have come to think of this way of eating as therapeutic nutrition: Some of us will just have to abstain from eating carbohydrate-rich foods—specifically, sugars, starchy vegetables, and grains—if we want to be relatively lean and healthy and stay that way. Understanding that simple fact, they say, can make this way of eating eminently sustainable. They believe this partly because of their clinical experience, and partly because considerable research indeed now demonstrates that this way of eating is inherently healthy. Slowly and steadily, conventional thinking about the causes of heart disease and the dietary triggers of chronic disease is shifting.

Many physicians, like Wolver, can sound like zealots or evangelists when they talk about these diets. A phrase I heard repeatedly in my interviews

---

led a generation of researchers to advocate calorie-restricted, low-fat diets for obesity, told me several times that "Atkins outdoes all others for weight loss" but it's not suitable for weight maintenance because "people tend to slip and let carbs back in."

for this book was that these doctors could not "unsee" what they had witnessed, both in themselves and in their patients. As more than one of these physicians told me, their discovery of a dietary means to prevent and treat obesity and diabetes—the disorders that overwhelm their practices—and one that was easy to follow, had made them excited again about practicing medicine.

Maybe evangelism is an appropriate response. A passionate doctor is not automatically a misguided one. Consider a story Wolver told me in July 2017. The previous February, she said, she received a phone call from a colleague who had just diagnosed diabetes in a twenty-four-year-old unmarried woman. This young woman's hemoglobin A1c—a measure of how well she could control her blood sugar and therefore the severity of her diabetes—was 10.1. Physicians consider levels above 6.5 to be diabetic. Over 10, according to American Diabetes Association guidelines, and the patient should be started promptly on insulin therapy.

"Do you think she'd ever get off insulin?" Wolver asked rhetorically. "Never. So my colleague said to me, 'I know you have a long waiting list, but can you see this patient? She's in my office, scared to death, crying.' I saw her the next morning. I explained to this young lady what she had to do, how she had to eat, and she started that day. I just saw her for her three-month follow-up. Her hemoglobin A1c was down to 6.1, no longer in the diabetes range. She had lost twenty-five pounds. When I told her she was no longer diabetic, she was crying. I called my colleague over, and **she** started crying. **I** was crying. I literally felt like I had cured cancer. This girl has her whole life in front of her, and it is not going to be spent on insulin, managing a chronic disease."

This was not a unique occurrence, a one-off, as skeptical critics refer to these experiences when they want to discredit them. In October 2017, more than one hundred Canadian physicians co-signed a letter to **HuffPost** publicly acknowledging that they personally follow LCHF/ketogenic regimens and that this is the eating pattern they now prescribe to their patients. "What we see in our clinics," these physicians wrote: "blood sugar values go down, blood pressure drops, chronic pain decreases or disappears, lipid profiles improve, inflammatory markers improve, energy increases, weight decreases, sleep is improved, IBS [irritable bowel syndrome] symptoms are lessened, etc. Medication is adjusted downward, or even eliminated, which reduces the side-effects for patients and the costs to society. The results we achieve with our patients are impressive and durable."

With the conventional dietary guidelines, they added, none of this happens: "Patients remain diabetic and still need medication, usually in increasing dosages over time. Don't we say that type 2 diabetes is a chronic and progressive disease? It doesn't have to be this way. It can actually be reversed or put into remission. Of the patients that we treat with a low-carb diet, most will be able to get off the majority or all of their medications."

These declarations, of course, come with critical caveats—as does Wolver's story and those of all the physicians and their conversion experiences. First, they are anecdotes, evidence only that these responses can happen when people abstain from carbohydrate-rich foods, not that they always or even almost always happen.

Second, they are incompatible with the conventional thinking on diet and health, which is why they are attacked as quackery. Not only do medical authorities, with the best of intentions, get appropriately nervous when mere MDs (let alone journalists like myself) start talking about reversing chronic diseases or putting these diseases into remission with unorthodox dietary approaches, but the way of eating that these physicians prescribe— one that allowed Wolver's young patient to lose twenty-five pounds in three months and put her diabetes into remission—one that this book will also recommend, clashes conspicuously with our widely held beliefs about healthy eating.

The very simple assumption underlying the LCHF/ketogenic diet is that it's the carbohydrate-rich foods we eat that make us unhealthy: both fat and sick. These are relatively new additions to human diets, so it shouldn't be a surprise that removing them can improve our health. Grains, whether whole or not, and even beans and legumes—the staples of a twenty-first-century conventionally "healthy" diet prescription—are to be avoided if at all possible. While naturally lean people may be able to eat these foods and remain lean and healthy, the rest of us may not. Of fruit, only berries, avocados, and olives are acceptable. And no matter how fat we might be, this way of eating does not advise us to consciously eat less or control our portions or count our calories or attend to how much is too much (or to take up running or go to spin classes). It advises us to eat when we are hungry and then eat to satiety, with the expectation that eating to satiety will now be relatively easy to accomplish.

More radical still, this way of eating is particularly, exceedingly fat-rich and tends to consist mostly of animal products (although, as I'll discuss, it doesn't have to be). It allows, even encourages, red meat, butter, and processed meats like bacon, and therefore animal fats and saturated fat. It can include copious green leafy vegetables but is not "mostly plants," nor in any conventional way "balanced." It commits the cardinal dietary sin of essentially excluding an entire food group.

This dietary approach—LCHF/ketogenic eating—is effectively identical to what Robert Atkins began prescribing in the 1960s. It is "Atkins redux," as the low-fat diet proponent and longtime Atkins foil Dean Ornish calls it. Atkins's prescription, in fact, was little different from the diet prescribed by the Brooklyn physician Herman Taller, whose 1961 book **Calories Don't Count** sold two million copies* and was described by a Harvard-trained nutritionist in the **Journal of the American Medical Association** as "a grave insult to the intelligent public." Taller learned of the diet from Alfred Pennington, who never wrote a book about it but used it to slim down obese executives at the DuPont Corporation in Delaware beginning in the late 1940s. Pennington published his results in medical journals, including the **New England Journal of Medicine,** and lectured about his work to a mostly positive reception at Harvard.

Pennington had learned about it from Blake Donaldson, a cardiologist in New York City who had worked in the 1920s with one of the founders of the American Heart Association and would prescribe it to his patients, almost twenty thousand of them, over the course of forty years. As a cardiologist, Donaldson may not have realized that he was rediscovering a nutritional approach to obesity that had been embraced by European medical authorities in the latter years of the nineteenth century,

* It was ghostwritten by the legendary sportswriter Roger Kahn, whose 1972 book **The Boys of Summer** is considered one of the best sports books ever written.

prompted by the publication of the first internationally best-selling diet book (technically a pamphlet), "Letter on Corpulence, Addressed to the Public," written by a London undertaker named William Banting, who reported that he lost fifty pounds by giving up starches, grains, and sugars. Banting, apparently unaware, was just repeating what the French gastronome Jean Anthelme Brillat-Savarin had written in 1825 in **The Physiology of Taste,** which would become perhaps **the** most famous book ever written about food and eating. After Brillat-Savarin concluded that grains and starches are fattening and that sugar makes it worse, his recommended diet for obesity was "more or less rigid abstinence" from those foods. This is the very advice that remains controversial today, the foundational core of the keto fad, and the simple idea that this book will flesh out.

The name continues to keep changing and the approach shifts subtly from year to year and from diet book to diet book largely because as physicians embrace it and conclude that it works—or stumble upon this particular reality themselves, unaware of its history, or find new ways of refining the basic idea—they write yet new diet books, with their minor variations on the theme, either to spread the word as widely as they can or to cash in (depending on your level of cynicism).

Despite the long and rich pedigree of this way of eating, academic authorities and the orthodox

still widely consider these LCHF/ketogenic variations, every last one of them, to border on quackery. In January 2018, just two months after the
publication of the aforementioned **HuffPost** letter, the supposedly authoritative annual diet review published by **U.S. News & World Report**
rated variations on these LCHF/ketogenic programs the least healthy imaginable—thirty-fifth
through fortieth of the forty diets reviewed. (The
publication has acted similarly in the past.) Only
Eco-Atkins (a vegetable-, vegetable-oil, and fish-
heavy version) and South Beach (similar) sneaked
into the top twenty-five, and the paleo diet tied for
thirty-second (alongside the raw food diet and just
below the acid-alkaline diet). The 2019 rankings
are more of the same.

To the physicians who now prescribe the LCHF/
ketogenic way of eating to their patients, what
their patients experience and their own eyewitness
testimony, what they cannot **unsee,** are far more
compelling than the fact that medical organizations and the kind of orthodox authorities enlisted by **U.S. News** to appraise diets still consider
LCHF/ketogenic eating much more likely to cause
long-term harm than any meaningful benefit.

For these physicians and their patients, the benefits are not only clear but also easy to quantify.
Patients undeniably get healthier. The number
of clinical trials supporting the benefits of these
diets has risen to near one hundred, if not more,

making it among the most rigorously tested dietary patterns in history. "This is not a fringe diet anymore. It's becoming mainstream" is how Robert Oh, a sports medicine and family medicine physician who is also a U.S. Army colonel, described it to me. Oh worked in the Office of the Surgeon General of the Army on an initiative to improve the health and readiness of troops and is now chief of the Department of Family Medicine at Madigan Army Medical Center outside Tacoma, Washington. "The best thing for me as a practicing physician," Oh said, "is that I can also share the stories of my patients with each other. I can say to one patient with type 2 diabetes, 'Look, I've got other patients exactly like you, and their labs have improved, and some are no longer on any medications.' And when other doctors see my patients, they're going to wonder how they got so healthy and ask what they did. And now they'll consider it for their patients. It's out there and spreading. Even the dietitians and authorities who are just blindly opposed to it can't stop it because it works."

Every time the World Health Organization or the U.S. Department of Agriculture or the United Kingdom's National Health Service or the American Heart Association proclaims in its dietary guidelines that a healthy diet **must** include fruits, beans, and grains (whole or not), that meats should be lean, fat should be avoided, and saturated fats should be replaced by polyunsaturated vegetable oils, it directly conflicts with these clinical trials and, more important, what these physicians are seeing daily in their clinics and their lives. It makes the job of these physicians, as they now see it, harder, but it doesn't deter them. It makes it harder for all of us who are not naturally lean and healthy to get there,* but it shouldn't deter us, either. From the perspective of these physicians, avoiding carbohydrates and replacing the calories with naturally occurring fats is indeed the therapeutic nutrition that their patients, and many of us, should be eating for life. As Paul Grewal, a New York City internal medicine specialist who says he has personally maintained a hundred-pound weight loss for eight years with LCHF/ketogenic eating, put it, "To be successfully reversing a disease and to be told not to do it or advise it to a patient is the height of absurdity."

Those of us engaged in this conflict, and particularly the physicians and dietitians on the front lines, believe that the advice we get from our public health, nutritional, and medical authorities is simply wrong, and that's why it fails, and that's why

---

* I include myself in this category, as the language suggests, because as a child I was what was then called "chubby," and my maximum weight as an adult was 240 pounds. Since I'm six foot two, that meant I had a body mass index (BMI) of 32, so I would technically have been considered obese, like everyone with a BMI over 30. I have also dieted, effectively, every day of my adult life. As I write this, I weigh approximately 210 pounds, which is, for me, a healthy weight.

so many people remain fat and diabetic, often miserable and burdened with medical bills. We have reached this conclusion based on evidence that we find compelling. We believe that an injustice is being perpetrated that has to be righted. Until we get these ideas understood and accepted—and tested as well as science will allow—not enough people are going to get the advice and counsel necessary to make a meaningful and sustainable difference in their own health and to curb the obesity and diabetes epidemics that are at large.

My hope is that this book will serve both as a manifesto for this nutrition revolution (to use an overworked but still appropriate term*) and as an instruction guide. The manifesto is necessary because meaningful change has to happen at a societal level as well as a personal one. That's why this book will discuss the mistakes made by the medical and nutritional authorities and the regrettable assumptions that we all came to embrace as a result. Ultimately we have to understand the simple chain of tragically bad science that led us into this situation. By doing so we can begin to fix what ails us.

I am presenting the instruction guide from

---

*For those who know their nutrition history, Atkins said much the same thing fifty years ago, which is why he put the word **revolution** in the title of his book, **Dr. Atkins' Diet Revolution.** I believe it was an appropriate response then, although foolhardy for a single physician like Atkins and perhaps ultimately counterproductive.

multiple perspectives. First, I'm synthesizing all that I've learned in twenty years as an investigative journalist reporting on and questioning the conventional wisdom on diet and chronic disease. (In the midst of unprecedented epidemics of obesity and diabetes, and the complete failure of our nutritional authorities and public health institutions and organizations to curb them, shouldn't that wisdom indeed be questioned?) I was fortunate when I began this investigation to be able to shadow clinical researchers like the Harvard University Medical School physician David Ludwig, who treated children with obesity at Boston Children's Hospital with what he calls a modified carbohydrate diet, and Eric Westman, who prescribed LCHF/ketogenic eating to his adult patients with obesity at his clinic in Durham, the same practice Sue Wolver would visit a decade later. These physician researchers and these experiences reminded me that what "most experts believe" in medicine is not always true, particularly when it comes to the treatment of obesity and the prevention of chronic disease. I was also fortunate that an MIT economist suggested to me that if I was writing about fat and weight, my research process had to include experimenting with the Atkins diet, upon which he had lost forty pounds; the father of one of his colleagues, he told me, had lost two hundred. I followed his advice, and the experience has informed (or biased, depending on your perspective) all that I've done since.

The advice and opinions are also informed by the physicians and dietitians I interviewed specifically for this book; they are listed in the references section and credited wherever appropriate in the text, footnotes, or endnotes. Their experience and observations inform everything I say. Evelyne Bourdua-Roy, a leader of this movement in Canada with a medical practice in the Montreal suburbs, summed up their thinking for me with a single line that she says she repeats to her overweight, obese, diabetic, and hypertensive patients. "I can give you pills," she says, "or I can teach you how to eat."

I also could not help but be influenced by the now thousands of people who have reached out to me, in the years since I first wrote about this subject in 2002 for **The New York Times Magazine,** to relate their experiences with this way of eating and thinking. These people had struggled their whole life with obesity and either won out over it or were still engaged in the struggle.

Finally, this book, despite its purpose as an instruction guide, includes no recipes or meal plans. I believe that learning how to think about how to eat, learning to understand what makes us fat and diabetic, means implicitly learning what to cook, how to order in a restaurant, and how to shop at the supermarket. Since my expertise does not in any way include cooking, please search out recipes and the necessary culinary guidance, which are now freely available on the Web and particularly at such invaluable sources as Dietdoctor.com, Diabetes.co .uk, and Ditchthecarbs.com. These sources will link you to others and to a world of cookbooks that will do a much better job of conveying what to cook than I ever could. My goal is to help each of us shed a century of tragic preconceptions about the nature of a healthy diet, to learn to ignore the bad advice we have been given, and to replace it with a way of thinking about diets, our weight, and our health that works. After that, the eating and the cooking should be easy.

1

The Basics

A brief lesson in the history of obesity research

On June 22, 1962, a Tufts University Medical School professor named Edwin Astwood tried and failed to correct how we think about the cause of obesity. We have been living with that failure ever since.

Astwood was presenting a counterargument to what had become since the end of the Second World War the dominant thinking among medical authorities and researchers on why we get fat. Astwood called this thinking "the conviction of the primacy of gluttony," by which he meant the unshakable belief that virtually all cases of obesity, child or adult, mild or extreme, are caused ultimately by the overconsumption of calories; that is, people get fat because they eat too much.

Astwood considered this belief system—for that's what it is—to be almost willfully naïve and perhaps the primary reason so little progress had been made in understanding obesity, let alone preventing and treating it. It is also the reason those who have the misfortune to suffer from obesity are held responsible for their condition. "Obesity is a disorder," he said in opening his presentation, "which, like venereal disease, is blamed upon the patient," the direct consequence of their failing.

Astwood was an endocrinologist; his medical expertise and the subject of his research were hormones and hormone-related disorders. The venue for his talk was the forty-fourth annual meeting of the Endocrine Society. Astwood was its president that year, and his talk, titled "The Heritage of Corpulence," was his presidential address. Astwood was also a member of the prestigious National Academy of Sciences. According to his NAS biographical essay, his peers considered him "a brilliant scientist" who had contributed more to our understanding of thyroid hormones and how they work than anyone alive. (He won the Lasker Award, considered one step below the Nobel Prize, for the thyroid work.) Of the young men and women who learned to do their medical research in Astwood's Boston-area laboratory, thirty-five would go on to become full professors by the time Astwood passed away in 1976. He was "not only driven by an insatiable curiosity," the NAS biography says of Astwood, "but by a curiosity that sought answers with willful determination."

Although Astwood was known among his friends and colleagues for having little interest in food or eating—he considered meals only "a necessary intervention in the day's activities solely for the purpose of bodily nutrition"—much of his laboratory work in the latter years of his research career was dedicated to understanding obesity, specifically the influence of hormones on fat accumulation and the use of fat to fuel our metabolism.

In the small world of 1960s-era obesity research, Astwood was something of a throwback to the pre–World War II years. While he had a profound understanding of the research literature on obesity and was a serious if not indeed brilliant scientist, he had been a physician also who treated patients in his clinic. In this he was like the physician researchers in Germany and Austria before the war who had dominated thinking on obesity and had also come to their conclusions on the nature of the obese condition by observing it closely in their human patients, taking their histories and coming to understand what they were going through and living with. Doctors would do that with any other disorder—why not do it with such a seemingly intractable disorder as obesity?

Many of the most influential of those prewar European authorities had become convinced that obesity must be the result of a hormonal or metabolic dysfunction, not **caused** by overeating, a concept that they recognized as circular logic. ("To attribute obesity to 'overeating,'" the Harvard nutritionist Jean Mayer had aptly commented eight years before Astwood's presentation, "is as meaningful as to account for alcoholism by ascribing it to 'overdrinking.'" It's saying the same thing in two different ways, at best describing the process, not explaining why it's happening.) Rather, it's somehow programmed into the very biology of the fat person, a disorder of fat accumulation and fat metabolism, these German and Austrian clinical researchers concluded. They believed, as Astwood came to believe, that obesity is neither a behavioral issue nor an eating disorder, not the result of how much we choose to eat consciously or unconsciously.

That German-Austrian research community had evaporated, beginning in 1933 with the rise of the Nazi Party. By the time the war was over, European thinking on obesity, grounded in decades of clinical experience and observation, had evaporated with it. The very lingua franca of medicine shifted from German prewar to English postwar. German-language medical literature was considered of little interest, even unreadable by the new generation of young American physicians and nutritionists, who repopulated the field and found the conventional, simplistic thinking on obesity all too easy to believe. With just a few exceptions, these newly minted experts weren't burdened with actually having to help obese patients achieve a relatively healthy weight for life. They were guided instead by a theory—technically, a hypothesis—that they believed in unconditionally. They believed the truth was obvious, which is always an impediment to making progress in any scientific endeavor.

Their truth was the subject of Astwood's presentation: a "conviction in the primacy of gluttony," the notion that obesity is almost invariably caused by eating too much, consuming more calories than we expend, and so is ultimately a behavioral or eating disorder. That conviction implied that the only meaningful difference between lean people and people who struggled with obesity is that the lean can control their food intake and hence their appetites—consume only as many calories as they expend—while people with obesity could not, or at least not once they started to get fat. The idea that the fat tissue of those who become obese might have some physiological drive to accumulate fat that the tissues of lean people don't, some subtle hormonal disruption, was dismissed by the authorities as nothing more than "lame excuses" (quoting the Mayo Clinic's leading 1960s-era obesity expert) for fat people not to do what came naturally to lean people—eat in moderation.

If anything, the supposedly learned postwar authorities came to consider obesity the result of a psychological defect, not a physiological one. They were not shy in stating that people got fat primarily because of "unresolved emotional conflicts" or because they had "turned toward food to relieve some of the nervous tensions of life." These authorities counseled those with obesity to embrace a lifetime of walking away from their meals still hungry, of semistarving themselves, ideally after consulting a psychiatrist first.

This is the thinking that Astwood hoped to overturn with his presidential address. He enumerated with elegance and occasional humor the reasons why obesity was surely a genetic disorder, which implied that it almost assuredly had to be a hormonal or endocrinological one. Yes, he acknowledged, this was the implication every time someone afflicted with obesity made a comment along the lines of "everything I eat turns to fat." It was anything but a lame excuse, according to Astwood; it was a reality. It was true, he said, not just for the kind of extreme obesity that he occasionally saw in patients in his practice, but for "the common or garden varieties . . . the kind that we see every day."

One thing that seemed to mystify Astwood was that there was nothing subtle about the evidence arguing for a genetic, and so hormonal, influence in obesity and fat accumulation. Obesity ran in families, Astwood said, as the authorities all agreed, but not because fat parents overfed their children. It did so because of a strong genetic component. Identical twins don't just have the same faces; they have identical body types. If one twin is obese, so almost assuredly will the other one be. Even the distribution of obesity in families suggested genetics were involved. Astwood told his audience about one of his patients who was twenty-four years old, five feet four inches tall, and weighed 457 pounds. This young man had seven siblings, three of whom also suffered from extreme obesity: "His brothers, aged 10, 15, and 21, weighed respectively 275, 380, and 340 pounds." The four other siblings "were of normal proportions."

This "looked more like the work of genes," said Astwood, not the "product of a groaning family board," an antiquated phrase that refers to a dining table overloaded with food. We know that genes determine stature and hair color, said Astwood, and they determine the size of our feet and a "growing list of metabolic derangements, so why can't heredity be credited with determining one's shape?" If we had doubts that this was the case, we only had to look at animals. "Consider the pig," he said: "His corpulence and gluttony resulted from man's artificial selection; selective breeding provided us with this hulk with his hoggish ways, and no one will convince me that his gourmandizing is provoked by parental oversolicitude."

A reasonable picture of how those genes might be expressing themselves, Astwood explained, had been worked out since the 1930s. A series of laboratory researchers had generated an enormous amount of information about how our bodies regulate the fat we store and the fat we use for energy. "To turn what is eaten into fat, to move it and to burn it requires dozens of enzymes and the processes are strongly influenced by a variety of hormones," he explained. Sex hormones clearly play a role in where fat is stored. Men and women, after all, tend to fatten differently: men above the waist, women below it. Thyroid hormones, adrenaline, and growth hormones all play a role in releasing fat from its depots, as does a hormone known as glucagon, secreted by the pancreas.

"The reverse process," Astwood said, "reincorporation of fat into the depots and the conversion of other food to fat, tends to be reduced by these hormones, but to be strongly promoted by insulin." All this demonstrated "what a complex role the endocrine system plays in the regulation of fat." An important clue to what might be happening, he added, is the fact that the numerous chronic disorders associated with obesity—"particularly those involving the arteries"—resemble those that come with diabetes so closely, it implies "a common defect in the two conditions."

Now imagine, Astwood suggested to his audience, what would happen if just one of these mechanisms went awry, impeding the release of fat from fat cells or promoting its storage. It was all too easy to imagine a slow, gradual accumulation of fat that could lead to extreme obesity if continued over years and decades. As the fat inexorably accumulated, a likely result would be what Astwood described as "internal starvation," as the body hoarded calories in fat cells that it would otherwise need for fuel, while simultaneously increasing the weight that had to be carried around, day in and day out, requiring the expenditure of more and more energy to move and fuel that bulk. In other words, the same subtle hormonal disruption that could cause fat to accumulate to excess would also make a fat person hungry while it was happening. This would be exacerbated by the advice given to the fat person from all sides: Eat less, exercise more. Starve yourself, if necessary. If the proposed treatment for a fat accumulation problem that itself caused internal starvation—that is, hunger—was to starve even more, we can imagine all too easily why it would fail, if not in the short run, certainly eventually.

"This theory," Astwood said, "would explain why dieting is so seldom effective and why most fat people are miserable when they fast. It would also take care of our friends, the psychiatrists, who find all kinds of preoccupation with food, which pervades dreams among patients who are obese. Which of us would not be preoccupied with thoughts of food if we were suffering from internal starvation? Add to the physical discomfort the emotional stresses of being fat, the taunts and teasing from the thin, the constant criticism, the accusations of gluttony and lack of 'will power,' and the constant guilt feelings, and we have reasons enough for the emotional disturbances which pre-occupy the psychiatrists."

Maybe the timing was bad, or the audience was wrong—a casualty of the silos in which medical research and medical practice tend to exist. Astwood gave his presentation right on the cusp of a revolution in the science of endocrinology. His comment about the intimate relationship between obesity and type 2 diabetes—the kind we're increasingly likely to get as we age and fatten—was remarkably prescient. It implied that the treatment and prevention of one would be very similar if not identical to that of the other. But he was talking to an audience of endocrinologists, who didn't treat the common form of obesity—"the kind we see every day," as Astwood had said. It wasn't their responsibility, and perhaps was not their interest, and in the early 1960s, obesity was still relatively uncommon compared with the epidemic confronting us today.

Back then, as Astwood implied, obesity treatment had become the purview primarily of psychiatrists and psychologists. These were the medical professionals charged with teaching fat people to get thin and supposedly elucidating our understanding of the disorder. They saw the obese and overweight, not surprisingly, from their own unique perspective and context, as clearly suffering from mental, emotional, and behavioral disorders. They found it easy to ignore a revolution in endocrinology, because that wasn't their area of study. (Nutritionists, as I will discuss, did the same.) They read different journals, attended different conferences, and were housed in different university and medical school departments. Even if the endocrinologists solved the problem, the psychiatrists and the psychologists might never know about it or might simply disagree, since they were diligently working to figure out how to get fat people to face up to their unresolved nervous tensions and eat less.

The fact is that by the time Astwood gave his presentation, the conviction in the primacy of gluttony had already won out. The world of obesity research back then was so small that a very few influential and well-placed individuals could and did determine what all the rest of them (and so us) would believe. "Obesity is a matter of balance—faulty balance of dietary intake and energy expenditure," they said repeatedly and with absolute assurance. It seemed so obvious that virtually all of us came to believe it unconditionally. Even some of the best and most empathic physicians of our era, such as Bernard Lown, a winner of the Nobel Peace Prize, bought in. He wrote in his classic book **The Lost Art of Healing,** subtitled **Practicing Compassion in Medicine,** that obesity is the result of "an innate maladaptive behavior," akin to "alcoholism, cigarette or drug addiction . . . absence of self-esteem, obsessive work habits, or simply a lack of joy in living." Even those suffering from obesity came to see their condition as their own fault.

By the 1970s, the idea that obesity is a hormonal disorder had effectively vanished from the learned discourse on the subject. The authorities, with only the rarest of exceptions, no longer even considered the possibility that we get fat because the hormones and enzymes that regulate the buildup of our fat stores and the breakdown and use of our fat for fuel are dysregulated in some of us and not in others, so that some of us fatten easily, accumulating excessive fat in our fat tissue or around our organs, and others don't. It is for this hormonal, physiological reason that some of us spend our lives fighting and losing a battle to remain lean, while others win it effortlessly.

Astwood's proposition and his theory, and the thinking of the prewar German and Austrian authorities, effectively disappeared. In 1973, after forty years of research had worked out the science of fat metabolism and storage in great detail, Hilde Bruch, the leading U.S. authority on childhood obesity, remarked on its absence. It was "amazing how little of this increasing awareness," she wrote, "is reflected in the clinical literature on obesity."

Today, nearly half a century later, this is still the case. While biochemistry and endocrinology textbooks diligently discuss the relevant details

of how hormones and enzymes regulate fat storage and metabolism and so imply that a subtle
disruption in these systems (particularly the hormone insulin) could easily cause human obesity,
just as Astwood had proposed, those very same
textbooks will omit this science entirely from the
discussions of obesity itself, as will textbooks dedicated entirely to obesity. Those discussions are still
dominated by the conviction of the primacy of
gluttony: It's the brain that makes us fat, and it
does so by manipulating how much we want to eat
and exercise. The absence of a competing theory
is remarkable, especially given the stakes and the
profound implications.

Imagine learned discussions of cancer—entire
books, even textbooks, written on the subject of
cause, cure, and prevention—that neglected to
mention, let alone discuss in detail, the physiological mechanisms that **directly** drive a tumor to
grow and a cancer cell to divide and multiply and
spread its progeny throughout the body. It would
never happen. Yet the direct equivalent did happen
in obesity research, and it has crippled our thinking on how we should deal with the disorder. The
physicians who are left with the job of treating an
ever-growing population of patients with obesity
and diabetes are expected to give their patients
variations on the same advice they would have
given in Astwood's day. And it continues to fail.

Also missing from these discussions has been

the direct and virtually unavoidable implications of this hormone-centric view of getting fat: the idea, or at least the possibility, that carbohydrates are uniquely fattening. Dietitians and nutritionists had accepted this as a given through the 1960s, but those researchers who thought of themselves as studying the causes of obesity failed to consider it a relevant piece of information. In 1963, Sir Stanley Davidson and Dr. Reginald Passmore wrote in the textbook **Human Nutrition and Dietetics,** the definitive source of nutritional wisdom for a generation of British medical practitioners, that "the intake of foods rich in carbohydrate should be drastically reduced since over-indulgence in such foods is the most common cause of obesity." They didn't understand yet why physiologically this was the case—it was just then being worked out in laboratories—but the fact seemed undeniable. That same year Passmore coauthored an article in the **British Journal of Nutrition** that began with the declaration: "Every woman knows that carbohydrate is fattening: this is a piece of common knowledge, which few nutritionists would dispute."

This observation resonated almost perfectly with what laboratory researchers were learning at the time about the hormonal orchestration of fat storage and fat metabolism. By excluding this thinking and its implications from mainstream medical practice—despite its being textbook medicine—

the authorities left it to the doctors themselves to do with it what they could, and they did. They found a way to eat that made it easy to achieve and maintain a healthy weight. Which brings us back to these "fad" diet books.

These books, written by doctors, sold so well not only because those of us who fatten easily have been desperate for answers but also because these carbohydrate-restricted diets—high in fat—provide for relatively quick weight loss and do so typically without hunger. The solutions provided in these books have simply been far more often right than what we've been hearing from the nutritional authorities. The advice works, for physiological and metabolic reasons that seem obvious. Yet the authorities, for reasons I'll discuss, have labored diligently to persuade us either that these diets won't work, or that we'll never follow them, or that if we do, they'll kill us prematurely. It's as though even trying this way of eating to see if it works were an affront to their expertise, which it is.

2

Fat People, Lean People

Fat people are not lean people who eat too much.

In the autumn of 2016, I was interviewed about fad diets for a BBC documentary. The host and interviewer was not a doctor but a highly respected University of Cambridge researcher who studies the genetics of obesity. (The academic articles on which he is an author have esoteric titles like "A Deletion in the Canine POMC Gene Is Associated with Weight and Appetite in Obesity-Prone Labrador Retriever Dogs.")

I assumed that the BBC producers wanted my thoughts because I was then (and may still be) the only journalist, historian, or scientist to write a detailed and critical history of obesity research: specifically, its convergence with nutrition, public health advocacy, and dietary guidelines. A few very good books had been written on the history of dieting and on nutrition research itself, but none with this greater context. (Please forgive the lack of humility.) My 2007 book, **Good Calories, Bad Calories** (published as **The Diet Delusion** in the UK), was the first that looked at this convergence, at the evolution of the thinking of clinicians and scientists on the cause of obesity and the chronic diseases associated with it—specifically diabetes, heart disease, cerebrovascular disease (stroke), cancer, and Alzheimer's—and at the implications for treating and preventing them by diet.

In writing that book, I had an advantage as a journalist that academics typically do not have: I could interview the players who ultimately changed the way we eat and defined our beliefs about the nature of a healthy diet (for better or worse). Along with reading the relevant available literature, from obscure academic papers to the published proceedings of the relevant conferences, I interviewed hundreds of clinicians, researchers, and public health administrators, some of whom were octogenarians or even older and had done their pertinent work or played their relevant roles half a century earlier.

I did this obsessive research because I wanted to know what was reliable knowledge about the nature of a healthy diet. Borrowing from the philosopher of science Robert Merton, I wanted to know if what we thought we knew was really so. I applied a historical perspective to this controversy because I believe that understanding that context is essential for evaluating and understanding the competing arguments and beliefs. Doesn't the concept of "knowing what you're talking about" literally require, after all, that you know the history of what you believe, of your assumptions, and of the competing belief systems and so the evidence on which they're based?* Because of this work, those researchers and physicians (as noted, a small but growing minority) who believed my interpretation of the science was likely to be at least mostly correct had come to consider me an authority, while those who didn't regarded me as a provocateur or a gadfly, occasionally even a quack. From the perspective of the latter, I am only a journalist meddling in medical and scientific issues.

The questions this Cambridge University geneticist wanted me to answer for the BBC were mostly variations on the theme of why people are drawn to fad diets. Why are doctors and diet books that push alternative ways to eat so eternally popular? Why are we so avid to read them? In all my years of research, I'd never actually thought to ask that

* This is how the Nobel laureate chemist Hans Krebs phrased this thought in a biography he wrote of his mentor, also a Nobel laureate, Otto Warburg: "True, students sometimes comment that because of the enormous amount of current knowledge they have to absorb, they have no time to read about the history of their field. But a knowledge of the historical development of a subject is often essential for a full understanding of its present-day situation." (Krebs and Schmid 1981.)

question, let alone answer it. Now the answer seemed suddenly obvious: Why not?

I'm talking specifically about those of us who are fatter than we'd like to be, overweight or obese— regrettably, a majority of the population these days. Most of the readers of diet books traditionally open them with the hope of learning how to control their weight, and today that can imply learning how to control their diabetes and hypertension, which so often accompany that excess weight.

The books that can be counted on to sell well are those that promise weight loss and weight control, ideally with little effort—"as if by magic," as Malcolm Gladwell described it in his 1998 **New Yorker** article on obesity and fad diets. This "as if by magic" concept is a critical one because it is what those who fatten easily are looking for. Rather than encouraging a lifetime of hunger and deprivation, diet books that sell well do so because they promise weight loss or the maintenance of a healthy weight in association with the full experience of good health: energy, mental clarity, improved sleep, freedom from the general ailments that come with aging and the stress of twenty-first-century lives. Readers tend to be those individuals—as the leading European authority on obesity, the University of Vienna endocrinologist Julius Bauer, described them in slightly more technical language back in 1941 (a bad year for European authorities)—who have "the compulsory tendency toward marked overweight due to abnormal accumulation of fat." This seems simple enough.

If we're still fat, with the tendency to get fatter, why wouldn't we look for alternative solutions? Wouldn't we be foolish not to? If we're already eating relatively healthy, if we already work to limit our portion sizes, if we go to the gym and maybe even count our daily steps on our wearable devices, and we're still fat or fatter than we consider ideal (not to mention maybe tired, sluggish, achy, sleeping poorly, and perpetually mired in a mental fog), then we're going to be attracted to popular diet books because the conventional approach is not working for us. Why not experiment with alternatives? Wouldn't any reasonable, thoughtful individual under those circumstances turn to different approaches to see if they work better?

I have seen little evidence that lean, healthy people can understand this thinking. The notable exceptions may be lean parents who have a child with obesity and must struggle to understand their child's experience. Perspective may not be everything, but it certainly plays a dominant role in how we come to understand the universe around us. "What you see is all there is," as the Nobel laureate behavioral psychologist Daniel Kahneman memorably put it. And the perspective of lean people—what they see—has been the determining factor in how the nutritional authorities have come to think about how **all of us** should eat. Those who are lean find it easy or at least relatively easy to control their weight. For this reason, they assume the rest of us can also do it.

Or, rather, they assume that we could if we were sufficiently motivated or had our priorities right. This line of thought leads quickly and directly to the not-so-subtle fat shaming that has been a forceful undercurrent throughout the last century of academic and medical thinking about obesity. (To read learned discussions of obesity and its treatment from the 1930s through the 1960s is to cringe with our twenty-first-century perspective at the shockingly biased, sexist, and degrading language used by these lean experts to explain why their not-lean patients stubbornly refused, once given the supposedly appropriate advice, to become lean themselves.) The same perspective problem exists for doctors. Those who are lean, and particularly those whose patients are also generally lean, have no reason to question the conventional thinking of the authorities. Whatever they're doing, it seems to be working for them and for their not-fat patients. They see little reason not to suppose that it works everywhere and for everyone. It's a natural assumption, but it's not a correct one.

This is why it's almost invariably lean people, or at least not-fat people, who counsel that all we have to do to achieve or maintain a healthy weight is to avoid "overeating," or to eat (as the lean journalist-turned-food activist Michael Pollan famously counsels) "not too much" or in "moderation," or (from the lean who think of themselves as particularly clever) to do "everything in moderation except moderation." They're implying, and they apparently believe, that this is sufficient to transform those of us who fatten easily into lean people like them who don't. (The same goes for exercise: Show me a lean marathon runner, and I'll show you someone who very likely believes that everyone would be lean if they all ran marathons, too.)

This is also why it's almost invariably lean, healthy people who advocate that we should eat effectively as we've been told to eat for the past fifty years—because it seems to work for them. Their logic is that surely those of us who are fat would be lean and healthy or become so if we did the same. At the very least, we wouldn't get fatter. So if we do get fatter by eating as they advise or if we have the misfortune to stay fat, it must be because we're not following their wise counsel, or because we just don't care. Hence the problem is our motivation and our priorities, and we should be ashamed.

Here's where the what-you-see-is-all-there-is problem is compounded by a lack of both curiosity and empathy. Those who disseminate this conventional thinking on weight control seem never to seriously question whether what they're assuming is actually true, whether maybe the world is full (and getting more so) of individuals who are overweight or obese who **do** eat healthy and in moderation, who **do** work out regularly, who **do** try diligently to eat "not too much." (Just as it may be full of lean people who do none of the above and yet remain resolutely and stubbornly lean.) Lack of curiosity and lack of empathy have always been defining characteristics of the official authorities on obesity and weight control and of most of the self-appointed (lean) authorities.

In fact, because those of us who fatten easily must work diligently at controlling our weight, even if we may fail, our waking hours (frequently our dreams, too, as Astwood noted) can often seem to be dominated by thoughts about what we eat or won't allow ourselves to eat, and how to moderate it. That's what we do. Many may eventually give up the fight, moving on to guilt or fatalism or both. Maybe apathy does set in because it seems hopeless: Obesity and diabetes seem to be our fate, no matter how conventionally healthy the foods we consume, no matter how meticulously we've eaten in moderation, no matter how faithfully we have followed the conventional advice.

Two points here are vitally important. The first is that the nutritional and academic authorities have failed us, and they and we should acknowledge that. Had they not failed us, we would, almost by definition, never have reached this point of epidemic obesity. That's the context of this discussion and all that follows. I believe it should be the context of every public discussion on obesity and weight control. **If the conventional thinking and advice worked, if eating less and exercising more were a meaningful solution to the problem of obesity and excess weight, we wouldn't be here. If the true explanation for why we get fat were that we take in more calories than we expend and the excess is stored as fat, we wouldn't be here.** So many more of us would be lean and healthy, and books like this one would not be necessary. The failure of that conventional thinking is the root of all the confusion about diet and health, the decades of controversy that the media likes to call the "diet wars," and the obesity and diabetes epidemics worldwide ("a slow-motion disaster," as the former director general of the World Health Organization recently called them).

The second point is fundamental to the first. It's the direct implication of this idea that we get fat because we eat too much, that obese people cannot balance the calories we consume with the calories we expend, but lean people can. It is, quite simply, the very root of the problem. It's relatively simple and should be obvious. It should be easily fixable. Indeed, despite my almost twenty years researching the history of this controversy and living in its trenches, I still can't quite wrap my head around the fact that this problem went uncorrected for so long. And yet it did, and so we have to understand it.

Despite decades of obesity research, and billions of dollars spent in the laboratory and on clinical trials, the bedrock fundamental concept underlying all nutrition and dietary advice is that fat and lean people are effectively identical physiologically, and that our bodies respond to what we eat the same way, except that the fat people at some point in their lives ate too much and expended too little energy and so became fat, while the lean people didn't. (The journalist Roxane Gay in **Hunger,** her memoir of living with extreme obesity, points out that even the very word **obese** comes from the Latin **obesus,** which means "having eaten until fat.")

Authorities will use sophisticated medical terminology to talk about why they believe obesity is a "complex multifactorial disorder." They'll do so, in part, so we might excuse their failure to make any meaningful progress in treating and preventing it over the decades. But the reason they have failed is because of what their thinking implies, and it is indefensible. Every time a health organization or a figure of authority states that obesity is caused by or results from taking in more calories than we expend, by overeating, they're basing it on this assumption: The **only meaningful difference** between people who stay lean and people who get fat is that lean people balance their intake to their expenditure and fat people don't, or at least they didn't while they were getting fat.

Here's the BBC Cambridge University geneticist making this point (perhaps without even realizing it), in the very first lines of a 2016 article: "At one level, obesity is clearly a problem of simple physics, a result of eating too much and not expending enough energy. The more complex question, however, is why do some people eat more than others?" The latter question may be more complex, but he's not asking why some people accumulate more fat on their bodies than others. Nor is he asking why some people happen to fatten easily and others don't, just as we might ask why some breeds of livestock—pigs, cattle, sheep (or even obesity-prone Labrador retrievers)—fatten easily and some don't. He's asking why we eat more, and therefore eat too much, assuming implicitly that we must and that's why we're fat.

In the late summer of 2018 **Nutrition Action,** a newsletter of the Center for Science in the Public Interest, published a Q&A with a leading obesity researcher at the National Institutes of Health, in which this expert stated that his theory about the cause of obesity and the obesity epidemic—the conventional thinking—is that our society pushes excess calories into the food system and then unsuspecting fat people eat them. "People who have obesity," he said helpfully, "are likely eating many more additional calories."

Worth noting is that this NIH authority was talking about the end result (so far) of a century of medical and nutritional research on one of the most intractable chronic disorders known to man, one that significantly increases our risk of falling victim to every major chronic disease. The explanation for the existence of this disorder, we're still being told, is that some of us just eat too much—we're not sufficiently vigilant to ward off these calories being pushed upon us.

"We think regulation of hunger and satiety is key" is how this was phrased recently in a remarkably candid comment to **The New York Times** by Cecelia Lindgren, an Oxford University professor who studies the genetics of endocrinology and metabolism. "There is food everywhere," she said. "If you are a little bit hungry and someone puts out a big plate of doughnuts at your meeting, who's going to reach for the doughnuts?" The constitutionally lean can refrain, she was implying, but those predisposed to be obese simply can't help themselves. Call it the "reach for the doughnuts" theory of obesity. Lindgren was proposing that genes might determine why some people in this modern doughnut-rich food environment just couldn't stop themselves from eating too much, and that's why they didn't deserve blame. But reaching for the doughnuts is still a conscious act, a behavior. It implies that willpower should be able to control it.

The unspoken proposition is that if researchers could only figure out how to induce those of us who eat too much to rein it in, curb our out-of-control appetites, eat smaller portions, and refrain from reaching for the doughnuts, we'd lose weight or not fatten to begin with. This, again, evokes implicit judgments about why we might fail should we have the misfortune to remain fat. It's not a failure in our bodies, not some hormonal or physiological phenomenon, that drove us (but not our lean friends or siblings) to amass fat. Rather it's some behavioral quirk, whether moral turpitude, lack of willpower, lack of vigilance, or the sin of gluttony and/or sloth. That's why we're still fat. It's not the expert advice or thinking that's misguided. It's us.

This blame-the-fat-person, look-who's-reaching-for-the-doughnuts thinking, the moral judgments and fat shaming, has always been embedded within this idea that obesity is caused ultimately by overeating. Here's one of the many areas in this controversy in which it helps to know the history. This fat-shaming implication was institutionalized as far back as the 1930s by the University of Michigan physician Louis Newburgh, who was largely responsible for convincing decades of physicians and obesity researchers that obesity is indeed caused by eating too much—"a perverted appetite" or a "lessened outflow of energy," as he put it—and **not** by some hormonal or physiological defect. Obesity, he and his colleague Margaret Woodwell Johnston wrote in 1930, is "always caused by an overabundant inflow of energy." The cause is never an "endocrine disturbance"—that is, hormones—that would manifest itself as a tendency to store calories as fat rather than burn those calories as fuel. By Newburgh's dictate, the cause is always some form of overeating.

This left open, though, the obvious question: What causes this overabundance? Or, rather, why don't fat people voluntarily curb their appetites, curb the overabundant inflow, and not get fat? Is it only a question of willpower? This too requires an explanation (just as the NIH authority in **Nutrition Action** still has to explain why some of us eat too much in this food-rich environment and others don't). Hence Newburgh, and all those who have come after him, transformed a physiological disorder into a character flaw. The overabundant inflow, said Newburgh, is the result of "various human weaknesses such as over-indulgence and ignorance." My suspicion, and I hope I'm not doing the man a disservice when he's no longer around to take offense, is that Newburgh's thinking was strongly influenced by the fact that he appears to have been pencil thin.

Even in cases that seemed obviously hormonal—

the pounds of fat often gained by women, for instance, when they pass through menopause or after a hysterectomy, the surgical removal of the uterus—Newburgh refused to concede an explanation other than overindulgence and weakness. Endocrinologists who studied this "well known" phenomenon in animals had concluded by the late 1920s that a critical role for female sex hormones—particularly estrogen—in the process of fat accumulation was implied. Secrete less estrogen, as women do during this phase of their lives or after a hysterectomy, and fat will accumulate. It happens to female animals. Maybe it should be no surprise it happens to female humans, too. So this, at least, must be hormonal. Not so, insisted Newburgh. It's **all** eating too much: "Probably she [the woman getting fatter as she goes through menopause] does not know or is but dimly aware that the candies she nibbles at the bridge parties which she so enjoys now that she is rested are adding their quota to her girth." Very scientific, that.

Time magazine memorably captured this thinking again in 1961 when it kicked off the apalling mistake of what became the low-fat diet movement with an influential cover story on the University of Minnesota nutritionist Ancel Keys. Just as Newburgh was central to disseminating the notion that the only meaningful difference between the fat and the lean is in their ability to control their appetite, Keys managed to convince medical authorities worldwide that we get heart disease because we eat too much fat or at least too much saturated fat. **Time**'s story on Keys and the evils of fat—both dietary and body fat—quoted the textbook **Harrison's Principles of Internal Medicine** referring to "the most common form of malnutrition" as "caloric excess or obesity," as though the two were one and the same. The **Time** article then observed that obesity in Puritan New England was seen as sinful, implying that perhaps it should still be, and quoted Keys saying, "Maybe if the idea got around again that obesity is immoral, the fat man would start to think."

The ridiculous implication, of course, was that if we did think about it (or if that self-indulgent, menopausal housewife did, rather than nibbling bonbons while she played bridge with her lady friends), we'd stop eating too much or at least stop eating immoderately; we'd control our portion sizes and our cravings and be lean. Our problem would be solved. Whether they know it or not, every doctor, every dietitian and physical trainer and friendly neighbor and sibling, every figure of authority who has ever counseled that we eat less and exercise more to lose weight, that we count our calories and so try to consume fewer than we expend, is wedded to this idea that the lean and the eventually-to-become-obese are physiologically identical; only their behavior sets them apart.

This belief system has dominated our thinking

on obesity since the 1950s, and we have to leave it behind. There are so many things wrong with this idea, things that were already known to be wrong in 1961 and even 1931, that it's hard to enumerate all of them. One of the most obvious problems with this thinking is that the logic is circular. Some very good clinical researchers pointed this out repeatedly in the mid-twentieth century, but these physicians-and-nutritionists-turned-moralists didn't seem to care. If we get fatter, more massive, we are clearly taking in more energy than we expend, and yes, the excess is stored as fat (although technically as fat and some muscle or lean tissue to support it and move it around as necessary). So we must be overeating during this fattening process. But that tells us nothing about the cause. Here's the circular logic:

Why do we get fat? Because we're overeating.

How do we know we're overeating? Because we're getting fatter.

And why are we getting fatter? Because we're overeating.

Logicians know this kind of round-and-round logic as tautology. It's saying the same thing in two different ways but offering no explanation for either. If we're getting fatter, it means our body mass is increasing, our energy stores are increasing, and so we are indeed taking in more energy—calories—than we expend. Okay, we're overeating. But by the same token, if we're getting taller we're

taking in more calories than we expend. But nobody would say we get taller because we overeat.
If we're getting richer, we're making more money
than we're spending. But nobody would say we
get rich because we overearn. That's clearly absurd,
even if overearning is what's happening as we get
rich, which it is—by definition. So why is this
kind of circular explanation considered acceptable
for obesity? It only appears to be an explanation.
It tells us nothing about causes.

The purpose of a hypothesis in science is to propose an explanation for what we observe, either
in nature or in the laboratory (ideally, a testable
hypothesis): Why did this happen and not that?
The more observations a hypothesis can explain
or the more phenomena it can predict, the better the explanation, the better the hypothesis. This
insistence that we get fat because we overeat is **not
even wrong,** as the legendary physicist Wolfgang
Pauli, a man with a gift for memorably pithy criticisms, might have put it. It explains nothing.

The counterargument, which I'm defending, is
Astwood's belief that those who fatten easily are
fundamentally, physiologically **and metabolically**
different from those who don't. This implies that
those of us who fatten easily can get fat on precisely the same food and even the same amount
on which lean people stay lean. We can't be told to
eat like lean and healthy people eat and expect that
advice to work, because we get fat eating like lean

and healthy people. Indeed, we get fat **and** hungry eating like lean and healthy people do. We need to eat differently. The question is how.

This observation about the physiological nature of obesity was made decades ago, perhaps centuries ago. The most conspicuous examples are animals (as Astwood noted with his "consider the pig" point) and the animal models of obesity that nutritionists and obesity researchers have studied since the late 1930s. Indeed, researchers would occasionally admit that it's clearly true about animals and animal models of obesity—that some animals get fat effectively independent of how much they eat and even when they eat no more than lean animals—but then somehow reject its relevance to humans on the basis that everyone knows that humans get fat because they eat too much. Their devotion to their energy balance thinking and to its implications was so great that they couldn't escape it.

Take, for instance, Jean Mayer, the most influential American nutritionist in the 1960s and into the '70s. Mayer started his research career at Harvard in the late 1940s and then moved on to become dean of Tufts University. The nutrition school at Tufts was later named after him. As a nutritionist, Mayer got some things right and many things wrong, as scientists often do, even the best of them. He spent the later years of his life arguing that people with obesity get that way because they don't exercise enough. Our current obsession with physical activity is largely rooted in Mayer's proselytizing in the 1970s. But at the beginning of his career in the 1950s, he studied a strain of obese mice. "These mice," he wrote, "will make fat out of their food under the most unlikely circumstances, even when half starved."

That's the nature of overweight and obesity. That's what it means to have a "compulsory tendency toward marked overweight due to abnormal accumulation of fat." Mayer's mice did not get fat by overeating. They got fat by eating. Half-starving them didn't make them lean. It only made them hungry and slightly less fat.

So let's redefine what we mean by obesity. People with obesity are not thin people who couldn't control their appetites (for whatever reason, psychological or neurobiological) and therefore ate too much. They're people whose bodies are trying to accumulate excess fat even when they're half-starved. The drive to accumulate fat is the problem, and it's the difference between the fat and the lean. The hunger and the cravings, and then the failures and the sins, as Astwood suggested, are the results.

This observation should be blindingly obvious to anyone who has ever had a weight problem, who fattens easily. Those who fatten easily are profoundly different from those who don't and may have been from the womb onward. Their physiology is different; their hormonal and metabolic responses to foods are different. Their bodies want to store calories as fat; the bodies of their lean friends don't. In George Bernard Shaw's play **Misalliance,** written in 1909–10, his character John Tarleton puts it this way: "It's constitutional. No matter how little you eat you put on flesh if you're made that way." Shaw, via Tarleton, may have been exaggerating slightly, but that's as good a way to capture the simplicity of the idea as any. If these people want to be relatively lean and healthy, if such a thing is possible, they have to eat differently. There may be foods they cannot eat. Foods that make them fat may not make their lean friends fat.

In 1977, in one of the more perverse episodes in the history of our ongoing discourse on obesity, a subcommittee of the U.S. Congress held a hearing in which the assembled congressional members listened to the leading academic experts of the day expound on the cause and treatment of obesity and its supposedly vital relationship to calories consumed and expended. The testimony left Henry Bellmon, a senator from Oklahoma, scratching his head, perhaps because Bellmon seemed to know what it was like to fatten easily and struggle with his weight. Maybe he was talking about himself. If not, it was surely a loved one who had opened his mind.

"I want to be sure we don't oversimplify," Bellmon said. "We make it sound like there is no problem for those of us who are overweight except to push back from the table sooner. But I watched Senator [Robert] Dole in the Senate dining room, a double dip of ice cream, a piece of blueberry pie, meat and potatoes, yet he stays as lean as a west Kansas coyote. Some of the rest of us who live on lettuce, cottage cheese and RyKrisp don't do nearly as well. Is there a difference in individuals as to how they utilize fuel?"

The experts in attendance acknowledged that they "constantly hear anecdotes of this type," but they could offer no other words of explanation. Their conviction in the primacy of gluttony didn't allow it. In fact, the evidence has always been clear, but it can't be reconciled with the notion that obesity is caused by eating too much and exercising too little.

Like Senator Bellmon, those of us who want to achieve and maintain a healthy weight can't afford to think about obesity as an energy balance problem. It gets us nowhere we haven't already been our entire lives. We have to think about it as a hormonal, metabolic, and physiological problem, perhaps akin to diabetes, as Astwood suggested. Some of us who don't seem to have it now are going to get it as we age. Some of us aren't. Some of us can load up on ice cream, pie, meat, and potatoes and stay lean as a west Kansas coyote; some of us can't.

But the foods we eat strongly influence the hormones responsible, as I'll discuss. That's textbook medicine. As such, the ubiquitous and seemingly obvious advice to eat "healthy," as the authorities invariably define it, is not relevant to all of us. The adverb **healthy** in that advice is a synonym for eating as the lean and healthy tend to do, but we are not them. We fatten easily; they don't. Doing as they do might surely be better for us than eating the standard Western fare of processed foods— "foodlike substances," as Michael Pollan memorably called them—and drinking sugary beverages (sodas, fruit juices, energy drinks, mocha latte cappuccinos) morning to night, but that's not good enough. It may also do harm or at least continue harm to be done. We have to eat differently because we are different.

3

Little Things Mean a Lot

People who get fat trap tiny amounts of fat in their
fat cells every day; not so people who stay lean.

Let's probe a little further into what it means to fatten easily. The authorities on obesity and chronic
disease, as I said, have characteristically lacked
both curiosity and empathy. This is another area in
which this absence conspicuously manifests itself.

Consider the implication of Ancel Keys's take
on this: If the fat man got fat by eating too much,
a reasonable question to ask, out of curiosity,
would be, How much is too much? Academics
who put themselves forth as experts on the science of obesity will often make dogmatic state-
ments in support of the conventional thinking
without digging any deeper into the implications.
Consider this statement by the Harvard professor of medicine Jerome Groopman, writing in
The New Yorker: "The importance of calories—if

energy gained exceeds output, the excess becomes fat—remains one of the few unchallengeable facts in the field of dietary science." Even if this were true—unchallengeable—wouldn't it be important to ask, How much energy is problematic? How many calories gained or not expended? What exactly are we talking about here?

This number is easy to calculate. The first instance I could find of this calculation having been done was in the very early twentieth century by the German diabetes specialist Carl von Noorden, from whom Louis Newburgh inherited his belief that obesity is caused by eating too much. Nutritionists had just recently become enamored of their ability to measure not just the calories people consume but also how many they expend. In the late 1860s German nutritionists had invented a new device, a **calorimeter,** that allowed them to do it. Measuring calorie intake and expenditure became the fashion among these researchers and, as a result, central to their thinking. At the time it was essentially all they had. Von Noorden estimated that we had to overeat (or underexpend) by only two hundred calories a day to put on seventeen pounds of fat in a single year. We had to increase our body mass by only two hundred extra calories a day, storing them in our fat tissue, to accomplish that feat. That's the equivalent of the calories in sixteen ounces of beer.

So, very simply put, if every day we drink one pint of beer or one large glass of juice too many, or we eat a handful of peanuts too much (maybe thirty peanuts), and if those calories are stored in our fat tissue, in a half-dozen years we'll go from being lean to being exceedingly obese. Von Noorden himself seemed reluctant to accuse people with obesity of ignorance or overindulgence. He used this calculation instead to suggest that such a small number of excess calories might slip into our diets without being noticed, even by the vigilant. This is the extreme-obesity-as-accidental-overeating theory, and it's a tough one to embrace.

Gaining seventeen pounds of fat in a year is quite a lot. Few of us will ever experience that kind of extreme weight gain, though, of course, pregnant women do. But what about more subtle amounts, the kind of fattening many of us do as we settle into adulthood—a couple of pounds a year? That will total twenty pounds in a decade or forty pounds in two decades. That will take us from relative leanness in our twenties to obesity in time for our midlife crisis. The mathematics is much the same as von Noorden's. Rather than two hundred calories and seventeen pounds in a year, our fat tissue has to accumulate just under twenty excess calories a day to amount to two pounds of excess fat every year. (Since a pound of fat is considered to be roughly equivalent to 3,500 calories, here's the equation: 3,500 calories/pound times two pounds divided by 365 days equals about 19.2 calories stored as fat each day.)

Fewer than twenty extra calories a day. That's three peanuts' worth of calories. It's the caloric equivalent of **half a teaspoon** of olive oil. So the man who eats 2,500 calories a day and burns off or excretes only 2,480 of them, with the last twenty calories never making it out of his fat tissue, is doomed to go from lean to obese in twenty years.

Here's another way to think about it, as the researchers studying fat metabolism (not to be confused with those who thought they were studying obesity) came to do in Astwood's era. During the course of the day, your fat cells take up hundreds, if not a thousand or more, calories of fat. They do this after a meal and hold on to those calories temporarily, then eventually release them back into the circulation to be used for fuel by the other cells in the body. This mobilization of stored fat takes place every day and every night. When you're asleep, you're burning the fat you stored during the day. So now imagine that almost all those calories taken up are later released, metabolized, or excreted, but **twenty remain behind** in the fat cells or maybe are mobilized but never used. So the liver packs them up and sends them back to the fat cells every day. If so, you're destined to become obese. We are talking here about twenty calories being spread out among the **tens of billions of fat cells** in the typical human body—literally infinitesimal amounts **per** fat cell.*

If we imagine this idea as a fat-trapping scenario, we can begin to understand why some of us might get fat despite our best efforts. For reasons unknown, some of us have fat tissue that holds on to twenty excess calories' worth of fat every day. Many hundreds if not a thousand or more calories of fat go into the fat tissue for temporary storage every day—virtually all the fat we consume, to be precise, not to mention the fat our liver cells might make de novo (from scratch) out of the carbohydrates we consume—but twenty don't come out. Who knows why that might happen (I'll get to that), but that's the excess. That's the too much. That's the difference between someone who stays lean and someone who doesn't, between those predisposed to be lean and those predisposed to be fat. **That's it.** That's what it means to fatten easily.

Here's the next somewhat obvious question: If that is what's happening, how would we know when we

---

* According to the latest edition of **Williams Textbook of Endocrinology,** a mildly obese person might have 70 billion fat cells each containing maybe 0.6 millionth of a gram of fat or a little more than one five-millionth of a calorie of fat. Divide the 20 calories of excess fat up among all these cells, and it means each cell has to take up about a third of a billionth of a calorie more fat every day or perhaps increase its fat store by roughly 0.006 percent every day.

get to those last couple of bites, those last couple of sips, the last fifth of a mile not walked, the last two hundred steps of the thousands we've taken that day, that we've crossed the threshold to fat storage, that we're now overeating or underexpending? How would our bodies know? If we're dealing with this kind of fat-trapping scenario, how can we possibly balance that out?

No guarantee exists that eating less will do it— certainly not 2,480 calories each day instead of 2,500, because we can't even tell the difference between those two amounts, but what about 2,300 or 2,000 calories? If we skip our midday snack, will that be enough? How do we know our fat cells won't still take up and retain those twenty extra calories a day? Those twenty calories are far less than 1 percent of all the calories we've likely eaten that day, less than 2 percent of the dietary fat we've likely consumed. Maybe our fat cells will hold on to that tiny bit of fat every day, even when we're half-starved. That's what the research in animals has always implied.

While von Noorden saw this simple algebra as a rationale for how people might unconsciously eat too much, other experts in the early decades of the twentieth century saw it as a reason to question the whole way of thinking. Eugene DuBois, for example, the leading authority on human metabolism in the United States in the 1920s and 1930s, suggested in his seminal textbook that this simple mathematics tossed the whole energy balance (gluttony and sloth) concept of body weight regulation into the realm of the absurd. Considering how exquisitely accurate the imbalance has to be to avoid obesity, how few calories actually have to be stored in excess as fat every day to become obese, to lead to tens of pounds of excess fat every decade, he said, "There is no stranger phenomen[on] than the maintenance of a constant body weight under marked variation in bodily activity and food consumption." (Another phrase used by physicists to describe this kind of problem is "spherically senseless," meaning it makes no sense no matter which way you look at it.)

Considering these tiny numbers, let's try changing our perspective and rephrasing the question. Rather than ask why some of us get excessively fat, what with all the copious food and drink we have available, perhaps the question should be why all of us don't. It's one thing to err on the side of undereating and go a little bit hungry all the time, but most of us don't. We eat to satiety. So why don't we all get fat? Certainly animals don't walk away from their plates hungry. Why don't they all get fat? Russell Wilder, the leading pre–World War II authority on obesity and diabetes at the Mayo Clinic, did ask precisely this question in 1930: "Why then do we not all grow fat?" After all, he wrote, "we continue to be protected against obesity, most of us, even though we hoodwink our appetite by various tricks, such as cocktails and wines with our meals. The whole artistry of cookery, in fact, is developed with the prime object of inducing us to eat more than we ought." (That was almost ninety years ago when Wilder said "most of us" are protected against obesity. Today he might have to say "some of us," but his point is still a good one.)

All discussions about obesity and overweight, about the epidemics, in terms of treatment and prevention, should start with an understanding of the absurdity of these tiny numbers and their implications. The authoritative discussions and guidelines, from Newburgh and Keys until today, rarely do so. They ignore them. In 1953 the British endocrinologist Raymond Greene, the leading authority of his era (and brother of the novelist Graham Greene), described this avoidance as already "an old trick in [the medical] profession"— that is, "the suppression of inconvenient evidence." Then he added the obvious: "Ignoring difficulties is a poor way of solving them." This is a lesson for all of us: If we don't actually quantify what precisely we're trying to explain, we don't have to fret over whether our preferred explanation fails to do it.

The numbers today make this issue even more of a challenge to the idea of overeating as a cause of obesity. The epidemic about which we hear so much, and about which I and others are so concerned, represents a gain in the average American's weight of twenty to twenty-five pounds over thirty to forty years, excess fat accumulation, on average, of half a dozen to a dozen calories a day. Now we're talking about the calories in a single almond or a single gummy bear or less than one-eighth of a teaspoon of olive oil stored per day. If you gained thirty pounds of fat between high school and your fiftieth birthday, this is the amount of calories you stored in your fat tissue every day that you didn't burn or excrete. This is the daily difference between you and one of your enviably still-lean friends. If anyone thinks they can balance out their intake and exercise to stop their fat tissue from accumulating **that**—half a dozen or a dozen calories a day—they have a far better imagination than do I or the nutrition/obesity experts who solve this quandary by pretending it doesn't exist.

One of the many extraordinary people I've met in the course of my work and research is a young man who grew up extremely obese in Southern California. He was mocked for his weight as a child. By his eighteenth birthday, he weighed upward of 380 pounds, which was the highest, he says, his scale would go. Clearly genes are involved, since his father was obese and he has an uncle who peaked at almost 800 pounds before getting weight-loss surgery, known technically as bariatric surgery. My friend is a tall kid, six foot three, which means his body mass index (BMI) was at least 47 when he was eighteen, well into the range defined today as **morbid** obesity. If he weighed only 180 pounds, his BMI would have been 22.5 and he'd be smack in the middle of the range that the experts consider healthy. He'd have been lean, a nice strapping young man.

So the difference for my friend between his morbid obesity and a healthy lean life was the two hundred excess pounds he gained in eighteen years. To simplify the calculation, let's assume that the excess was all fat. (It would not have been; about a third of it would have been muscle, technically lean tissue, but the point is the same.) Accumulating that much excess fat in that time frame means that on average his fat tissue took up from his diet and then held on to one hundred excess calories of fat every day. That's the equivalent of the fat in a tablespoon of butter or a little more than a tablespoon of olive oil. It's the calories in less than a fifth of a McDonald's Quarter Pounder with cheese.

Now we have to believe that this young man got morbidly obese because he ate a fifth of a cheeseburger more than he should have every day, or maybe half a cheeseburger, if we're going to take into account the calories used up in digestion and absorption of the food and the fact that he's a lot heavier than his friends are and so his body expends more energy just by virtue of that. We are supposed to believe that had he shown the proper awareness of portion control and walked away before finishing that Quarter Pounder, he'd have been lean. Had he done so, not only would he not be morbidly obese, the argument goes, he would have avoided all the shaming and ridicule he faced through most of his childhood and adolescence.

Anyone who can sincerely believe that this simple exercise in very modest portion control was all this young man needed to do to remain lean is, to my mind, delusional. Rather than be morbidly obese and ridiculed for it, all he had to do was walk away from his meals having eaten "not too much," remaining just a little bit hungry. Let me add that my friend told me that never in his childhood or adolescence could he remember being satiated—in other words, he **always** walked away from his meals wishing he could eat more. What this young man did that his lean friends did not was accumulate fat all too easily.

When we assume that people who get fat are merely lean people who ate too much, we do them a terrible injustice. Considering the burden that obesity is to those who suffer from it, why would they not do the little things necessary to fix the problem if little things were indeed the cause, as the calorie model implies? I have an acquaintance in Texas who weighed 280 pounds at her peak. She was in her late forties when I first heard from her. "I'm an educated, successful professional with a happy marriage" was how she described herself. "I'm not supposed to be a failure. But the shame of fat has hung around my neck for nearly my whole life. For most of my life, if asked to describe myself, the very first word that would come to my mind was **fat.** Not any of the other ways a person might typically describe themselves: **female, daughter, sister, American, wife, professional, tall, blond, 48,** [fill in political affiliation], etc. Nope. My number-one descriptor, the thing that colored everything and defined so much of my life: **fat.**"

This is what happens when we assume that a disorder or a disease like obesity is caused by something as meaningless as caloric imbalance. It puts the shame on the person who suffers from it, who couldn't stop herself from being obese, from accumulating a tiny amount of fat daily, rather than on the community of authorities who have failed so conspicuously to understand it.

4

Side Effects

Chasing a calorie deficit is a fool's game.*

If fattening implies, as it does, that we merely accumulate every day in our fat cells some tiny percentage of all the fat calories we consume (plus whatever carbohydrates happen to be converted into fat), what **are** we expected to do about it? Isn't that the critical, key, obvious question?

The orthodox thinking is that we eat less—exercise portion control, maybe leave the table hungry. By a variation on the same mathematics that von Noorden used more than a century ago, dietitians will patiently explain to those who are overweight or obese that if they eat five hundred calories a day fewer than they prefer,

---

* This particularly colorful phrasing of the problem is not mine but that of Ken Berry, a family physician in rural Tennessee and author of the book **Lies My Doctor Told Me.**

or are currently consuming, they will lose a pound a week. One thousand calories fewer will mean two pounds. They will advise women to eat 1,200 to 1,500 calories total each day and men to target between 1,500 and 1,800, as the website of the National Heart, Lung, and Blood Institute now does.

The thinking is that if we cut back sufficiently on how much we eat, surely we will get that excess fat out of our bodies, regardless of how seemingly trivial the overeating might have been that produced that fat. In the 1960s and '70s these calorie-restricted diets were often known even in the research literature as semistarvation diets. I'm going to use that terminology because it is entirely appropriate.

This assumption that people will lose weight if they are starved sufficiently is certainly true. This is one reason clinical researchers and physicians from Newburgh onward were so convinced that we get fat because we eat too much. Cut back enough on the calories a fat person is allowed to eat, and the result is a less fat person. But as the Harvard psychologist William Sheldon observed in the late 1940s, starving a fat man (an endomorph, in his terminology) doesn't actually turn him into a lean man (an ectomorph) or a muscular, athletic one (a mesomorph) any more than starving a mastiff turns it into a collie or a greyhound. For the dogs, you get an emaciated mastiff. For the humans, an emaciated fat man.

So this thinking, too, has some serious problems that have to be ignored to embrace it. If you put a lean person on a semistarvation diet, you also get a less fat person—actually, an emaciated lean person. Starving or semistarving a growing child will result in an emaciated child whose growth is stunted, but no authority would ever assume, let alone state publicly, that children grow because they eat more than they expend. At least I hope not. Yet that's always been considered the reasonable interpretation of the starve-a-fat-man observation. The important question, however, is why it is that some of us have to be chronically starved or semistarved—exercise portion control and be hungry for a lifetime—to be lean, or at least leaner, and others don't. This is another question that is rarely asked.

Ultimately, the question that can be easily answered and that certainly should be asked of anyone who suggests we can regain a healthy weight by eating less (let alone that this way is ordained by the laws of physics) is, At what cost? What are the side effects? What are the negative **sequelae,** as doctors with a fondness for Latin terminology would call them?

We care about the side effects of any drug therapy for a disease. Does it give us headaches, make us drowsy or dizzy? Do we get abdominal pains or cramps, nausea, and vomiting? Diarrhea? Erectile dysfunction? If we take a drug to lower our cholesterol and it makes our muscles ache unbearably, we're going to find another way or at least another drug to do that job. So what about dietary therapy, and eating less specifically?

Imagine that we voluntarily decide to accept, as Keys suggested and Newburgh implied, that eating less is indeed an absolute requirement for weight loss and then continued weight maintenance. We aim to cut back sufficiently so that we become relatively lean and stay that way for a lifetime. What might that entail? What are the common side effects? What will we have to endure?

The authorities know the answer to this question, which may be why they rarely if ever ask it. The relevant seminal study, the one that has stood for decades as incontrovertible evidence (such as there is in nutrition research), was conducted in the early 1940s by Ancel Keys. He and his colleagues then wrote a two-volume tome, nearly 1,400 pages total, about all that they had learned. The title, **The Biology of Human Starvation,** immediately tells us a little about what he did and what the experience must have been like for the experimental subjects, who provided the answer to the question of what happens if we try to live with the kind of caloric deprivation that the authorities often argue is necessary for significant

weight loss. Put simply, we get hungry, exceedingly so. "The best definition of food deficiency,"
as Keys and his colleagues wrote in **The Biology
of Human Starvation,** "is to be found in the consequences of it."

In the early years of World War II, Keys and his
University of Minnesota collaborators enlisted
three dozen conscientious objectors for the experiment. Most of these young men were lean; a
few were heavy, at least by the standards of that
considerably leaner era. Keys and his collaborators fed them roughly 1,600 calories daily of what
would today be considered a very healthy if very
boring diet: "whole-wheat bread, potatoes, cereals
and considerable amounts of turnips and cabbage"
with "token amounts" of meat and dairy. It was
a low-fat diet, as nutritionists would call it, low
in saturated fat surely, so it was right in line with
the dietary guidelines of most twenty-first-century
health organizations. The calorie level would put
it well within the range recommended for weight
loss today.

For the first twelve weeks, the men lost an average of a pound of body fat a week, but this slowed
to a quarter pound weekly for the next twelve,
despite the continued deprivation. In total, that's
an average of fifteen pounds of fat shed over almost half a year. Not bad, although certainly not
all that great (keeping in mind that these men did
not have that much excess weight to lose). This,

however, was not their only response. The men felt continually cold. Their metabolism slowed. Their hair fell out. They lost their libidos. They threw tantrums and thought obsessively about food, day and night. "Semi-starvation neurosis," the Minnesota researchers called it. Four developed "character neurosis," which was more severe. Two of those had breakdowns, one with "weeping, talk of suicide and threats of violence." He was committed to a psychiatric ward. The "personality deterioration" of the other "culminated in two attempts at self-mutilation." The first time he nearly cut off the tip of one finger with an ax. When that didn't get him released from the study, he "accidentally" chopped off three fingers.

That's quite a price to pay for eating a healthy, mostly plant, whole-food, low-fat diet of 1,600 calories a day. When these men were allowed at the end of the study to eat to satiety, they consumed prodigious amounts of food—up to 10,000 calories a day. They regained weight and fat remarkably fast. After twenty weeks of recovery, they averaged 50 percent fatter than when they started—they had "post-starvation obesity," as Keys and his colleagues called it. Many of us have been there. We can relate.

So we know that lean, healthy people can't live with this kind of calorie restriction, not if they have any choice. Why expect a fat person to do it? In fact, you can ask any lean friends you might have, in all seriousness, what they would do if their goal in life or just for a single day was actually to make themselves hungry, to "build up an appetite" and keep it up. Tell them to imagine that they are invited to a feast that night, one with the best food they'll ever eat, course after course after course. Tell them their goal is to come hungry and bring an appetite. Ask them what they'd do to make sure that happens. I'm willing to bet that they'll suggest that they start by eating less during the day, skipping snacks, and reducing their portion sizes when they do eat, and they'll probably also decide that exercising will help, or exercising more—going for longer walks or hikes, burning more calories on the elliptical machine at the gym. In short, eating less, exercising more.

This again should tell us that we have to rethink our approach to preventing and curing obesity, that we need a different paradigm to understand how we get fat and how to lose that fat. The very same things that any reasonably sane lean person would do to build up an appetite—that is, to get hungry and stay hungry—happen to be the very same two things that we tell those who have excess fat to do to lose weight. And these are direct implications of the idea that people with obesity begin life just like lean people, then eat too much, that obesity is an energy balance problem rather than a hormonal one.

Of course, if those of us who are fat did try to

subsist on, say, 1,600 calories every day but failed to sustain it—because we were, well, continuously, chronically, cut-our-fingers-off-to-escape-it hungry, just as lean people would be—if we failed to maintain the necessary monitoring of our portion sizes and the daily exercising along with it, and now our prestarvation obesity manifested as poststarvation obesity, we would get blamed for lacking willpower. We would get blamed for committing the sins of gluttony and sloth, ignorance and self-indulgence. We would be told we were just not thinking enough to realize we should be eating not too much, or at least should have been eating not too much all along. Many responses come to mind, but they are, regrettably, unprintable.

5

The Critical **If**

The problem is in our bodies, not in our brains.

Hard as this may be to imagine, I'd like to suggest that the authorities, being all too human, just got it wrong—virtually all of them, from von Noorden in the 1900s to Newburgh in the 1930s and '40s and all those who followed. (Although as Malcolm Gladwell wrote in his 1998 **New Yorker** article about obesity and the obesity epidemic, considering how often the medical orthodoxy is mistaken, this should never be that hard to imagine.) Gluttony and sloth and overeating and eating too much and sedentary behavior and physical inactivity and even overindulgence and ignorance (or unresolved nervous tensions) are easy answers for why so many of us get fat, but they're wrong.*

* A few of the very best obesity researchers knew that they had made effectively no progress in understanding obesity. Jules Hirsch

They sound reasonable, so we, too, fall for them, but they are wrong. The authors of fad diet books have been trying to tell us so for decades. Some got it wrong, as I said, but many got it mostly right because the mostly right solution worked.

Why the authorities would make such an extraordinary mistake is more obvious in retrospect than it must have been at the time. The gist of it is that they thought about the problem (and still do) from a perspective that seems obvious (if you're a lean person) and just happens to be more than a bit deceptive. Not only did they think it meaningful that they could starve people and emaciate them, but they could all too easily imagine fat men like Shakespeare's Falstaff, with his gluttonous appetites for food and drink—"He hath eaten me out of house and home; he hath put all my substance into that fat belly of his," says Mistress Quickly about Falstaff in **Henry IV (Part 2)**—and

---

of Rockefeller University, for instance, whom **The Washington Post** once described as having "helped reframe the modern understanding of obesity," told me in 2002, shortly before he retired, that he considered his career an abject failure. After almost forty years of research, he could no more explain how people get fat to begin with than he could explain how they can lose weight and keep it off once they do. They both remained mysteries to him. "I've been working on this for a hell of a long time," he said. "You'd think I would have gotten a little further along with it." Four years later Hirsch won a lifetime achievement award from the Obesity Society. I'm surprised he didn't politely turn it down. (Langer 2015.)

then assume that if Falstaff got fat for living im-moderately, so must we all.*

But the **if** buried in that line of reasoning is an absolutely critical one. Even in Falstaff's case, we don't know **if** his gluttony caused his obesity or vice versa. Growing children will also tend to eat us out of house and home. (I shudder to think how much my voracious eleven-year-old will cost to feed as he's growing through puberty and ad-olescence.) They do it because they're growing. So maybe adults with growing bellies do so for similar reasons.

When my two sons were younger, we were fans of a series of humorous French children's books about young Nicholas (Nicolas, in the original French versions) and his schoolmates. Nicholas has a friend named Alec (Alceste), who is "fat and he eats all the time." When Alec isn't eating, he's hungry. He is invariably pulling the remains of a croissant or a pastry from his pockets to eat be-tween meals or even between snacks. He's all too ready to abandon the latest hijinks to ensure he makes it home promptly for dinner. But never in

---

* This thinking is so broadly accepted that even the Princeton phi-losopher and animal rights activist Peter Singer (writing with Jim Mason) uses it to argue that obesity is unethical. Aside from wasting food (and so the lives of animals) merely to accumulate body fat, he says, "if I choose to overeat and develop obesity-related health problems that require medical care, other people will probably have to bear some of the cost." (Singer and Mason 2006.)

the books does the author, René Goscinny, opine on the possibility that Alec is fat **because** he's hungry and eats all the time. Perhaps he eats all the time because his body, unlike that of Nicholas and their numerous other lean friends, is singularly dedicated to accumulating fat. Perhaps his hunger is the result, not the cause, of a "compulsory tendency to marked overweight and accumulation of fat." This is the point Astwood was making. Hunger is a response, not a cause.

This is a fundamentally different way of looking at the problem of obesity and why we accumulate fat, and if we want to end the obesity epidemic and deal successfully with our own weight problems, we will have to take it seriously. We will have to learn a different way to eat as well (as the low-carb and keto diet doctors have been saying for a while now). In the mid-twentieth century, many of the leading figures in obesity research—like Julius Bauer at the University of Vienna and Russell Wilder of the Mayo Clinic, who actually studied and treated patients with obesity and who thought critically about the problem of human fatness without preconceptions—had come to accept or at least seriously consider the possibility that the seemingly obvious explanation, the conventional wisdom, for the relationship among hunger, eating, and excess fat, that the first two cause the latter, gets the cause and effects backward.

Rather, it's the drive to accumulate fat, rather

than use it for fuel, that leads to the hunger and any seemingly excessive eating that occurs. They found this explanation compelling. These authorities had to struggle with it, though, because even they had been indoctrinated with the thinking about calories—the gluttony conviction.

These authorities were legitimately curious about the subject they were studying. They asked questions about the process of fattening that might shed important light on the problem. This was how Astwood was thinking. Why, for instance, do men and women fatten differently, and in very different places? Why do boys gain muscle and lose fat when they go through puberty while girls gain fat and do so in specific places (hips, buttocks, breasts)? Why do women gain fat as they go through menopause, the experience Newburgh and his followers wrote off to bonbons, bridge parties, and self-indulgence? Why do people get fat in some places (double chins, love handles) and not others? What about fatty tumors known as lipomas? Why do these benign fat deposits hold on to their fat even during starvation?

These kinds of questions, they concluded, could be reasonably answered only by postulating hormonal and enzyme-related explanations for fat accumulation and obesity. Surely how much people ate and exercised said nothing about these kinds of questions. If I have a pot belly but my legs are skinny as bean poles (as with a significant proportion of men over a certain age in America), it seems clear that the amount of calories I eat and expend can tell me nothing about why. It seems hormones must play a key role in fat accumulation, as they do most other processes in the human body, and that a subtle shift of these hormonal mechanisms (which includes the enzymes and receptor molecules that can be thought of as the cellular antennae that receive and respond to these hormonal signals), whether globally or locally, could explain human obesity and these questions of localized fat accumulation. That in turn implies, as Astwood had suggested, that maybe **all** relevant questions about fat accumulation and obesity demand or require these kinds of hormonal and enzyme-related explanations.

Ultimately, these pre–World War II physician researchers were thinking about the problem of excess fat from the perspective of first principles. Rather than asking why fat people eat so much or exercise so little (without even knowing how much of either they do, as is the common state of affairs), they asked why these people accumulate so much fat, and why they accumulate it when and where they do. What regulates the process of fat accumulation? Why is fat trapped in our fat tissue—or around our organs or in our livers, as is all too common these days, and dangerously so— and not used for fuel? Lean people burn fat for fuel. Why do those of us who are fat keep so much of it stored away? Why do some of us fatten easily while others don't?*

By the early 1960s, when Keys was hoping to shame fat men into thinking about their immoral behaviors, decades of very good scientists had already gone a long way toward answering these questions. This was what Astwood, the endocrinologist, considered relevant and Keys and his nutritional colleagues did not. Researchers—physiologists, notably, and so not physicians or nutritionists and certainly not psychiatrists or psychologists— had discovered that the storage of fat in fat cells and the liberation of that fat from storage and its use for fuel (oxidation, in the lingo) wasn't in any way the simplistic process that was implied then and is implied still by the nutritional authorities.

Columbia University's Hilde Bruch, who was **the** leading mid-twentieth-century authority on childhood obesity, understood this and waxed indignant about it in a book she wrote in 1957 called **The Importance of Overweight,** which should still be required reading for anyone interested in understanding obesity. Bruch said that when she started to study obesity in children in the late 1930s, her medical colleagues would often ask her how she could "possibly want to work with

---

* One key to making sense of the universe—i.e., doing good science— is knowing that the answers we get are dependent entirely on the questions we ask, so we'd better be asking the right questions before we conclude we got the right answers.

such dull and uninteresting cases." Her patients complained to her that their previous physicians had been uninterested in their situation or worse. "Quite often patients had recognized more than just lack of interest," Bruch wrote; "they had felt offended by a condescending or sometimes frankly punitive and condemning attitude."

Bruch herself was mystified, specifically, by the absence of interest by the researchers (those who had taken on the obligation of understanding obesity) in this process of fat accumulation. "Looking at obesity without preconceived ideas," she wrote, "one would assume that the main trend of research should be directed toward an examination of abnormalities of the fat metabolism, since by definition excessive accumulation of fat is the underlying abnormality. It so happens that this is the area in which the least work has been done." She added, "As long as it was not known how the body builds up and breaks down its fat deposit, the ignorance was glossed over by simply stating that food taken in excess of body needs was stored and deposited in the fat cells, the way potatoes are put into a bag. Obviously, this is not so."

Bruch understood this for many reasons, but I'm going to suggest here that she did so largely because she was a working pediatrician; she not only studied obesity in children—at Columbia, where she had opened the first pediatric obesity clinic in the United States—but she treated children who were obese, although with little success. These kids were not statistics to Bruch, numbers from a survey or answers from a questionnaire about what they might be eating and how much they might be exercising. They were her patients. She talked to them and interviewed them; she spent time with their parents and interviewed **them.** In doing so, she learned about both the compulsion to get fat **and** the compulsion to eat that might go with it.

Bruch also followed her young patients as they grew into young adults. As Bruch told it, she was initially impressed by how easily these kids lost weight when she first gained their cooperation. But by 1957 she was more impressed with how quickly they gained the weight back, "the tenacity with which they maintain their weight at an individually characteristic high level." So she concluded that "overeating, though it is observed with great regularity, is not the cause of obesity; it is a symptom of an underlying disturbance. . . . Food, of course, is essential for obesity—but so is it for the maintenance of life in general. The **need** for overeating and the **changes** in weight regulation and fat storage are the essential disturbances."

By 1957, as Bruch wrote in her book, researchers were coming to understand many of the ways that hormones and their cellular targets, enzymes, work to orchestrate the use of fat in our bodies— how, where, and when it gets stored and is then liberated back into the bloodstream for its use as fuel. For those like Bruch and Astwood who paid attention to this literature, it was all too easy to imagine how this complex biological system could somehow be out of balance in obesity, disturbed by some element of our modern world, such that we accumulate excessive amounts of fat in our fat cells (and maybe in and around our organs, too) in a way that is little influenced by how much we eat.

Researchers studying fat accumulation in animals would note how fat cells and the animals themselves could accumulate fat or mobilize it and burn it for fuel "without regard to the nutritional state of the animal," as though how much or how frequently the animal ate was irrelevant to whether it was using up its fat stores or building them up. As Jean Mayer, the then-Harvard nutritionist, would say about his laboratory mice, they turned food into fat even when half starved. Why not humans, too?

And if humans do, here's the obvious critical question: Can this fat-storage problem be fixed? Can we change the way we eat so that it no longer happens and the bodies of people with obesity work like those of lean people?

6

Targeted Solutions

The ideal diet works "as if by magic"
because it corrects the diet.

**It is in vain to speak of cures, or think
of remedies, until such time as we have
considered of the causes . . . and the
common experience of others confirms
that those cures must be imperfect, lame,
and to no purpose, wherein the causes have
not first been searched.**

—ROBERT BURTON, quoting Galen in
The Anatomy of Melancholy, 1638

The establishment authorities and the diet doctors
agree that the kind of diet we're discussing has to
be sustained—and sustainable—for a lifetime, or it
won't work for a lifetime. That's why the word **diet**
is inappropriate to refer to what has to be a lifelong

change in how or what we eat. **Lifestyle** is the preferred term, or **eating pattern.** It's why I refer to LCHF/ketogenic **eating** rather than LCHF/ketogenic **diets.** This also seems simple enough, and it's based on very simple logic. Diets work when we change what or how much we eat and it fixes what ails us. If we fall off the diet, it means we're going back to however or whatever we were eating that caused or exacerbated our problems. We'd be foolish to think the result is going to be any different than it ever was.

Here's a simple example of this logic: I have a corn allergy. I get various kinds of gastrointestinal (GI) distress from eating it. If I don't want the GI issues, I don't eat corn, and I do my best to avoid packaged or prepared foods that include corn products among the ingredients. I learned to do this in my childhood, and I continue to do it. We could say I'm on a corn-free diet, and I know that if I add corn back, I'm going to have the same problems I always did. Hence sustaining a corn-free lifestyle is easy for me, and sustainability isn't an issue. I just do it. My avoidance of corn is life-long because it has to be.

Slightly less obvious but nonetheless true is that all reasonable dietary approaches assume a hypothesis, implicitly or explicitly, about the cause of the problem that the diet is supposed to fix. If proponents of vegan and vegetarian eating are right about its health benefits (not to be confused with the ethical, moral, and environmental issues raised by eating animals), then meat and animal products are a root cause of our major food-related ailments, and avoiding meat and animal products will make us healthy or at least significantly healthier. When nutritional authorities tell us plant-based eating is the healthiest way to eat, they're hypothesizing that plant-based foods are better for us than animal products and that the latter are harmful, at least in comparison. But if we switch to plant-based eating and remain fat and/or diabetic anyway, or if we've been eating vegetarian or vegan all along, or mostly plants, and have become fat and/or diabetic, then it's likely that meat and animal products are not our particular problem, or at least not the principal problem, and it behooves us to correctly identify what is.

The hypothesis underlying the conventional wisdom on food and weight, as we've been discussing, is that we get fat by overeating and so the route to getting lean is undereating. Diets that work, as the authorities will say, are those that reduce calories and make us eat less. "All diets that result in weight loss do so on one basis and one basis only: they reduce total calorie intake" is how this is stated unconditionally in the most recent edition (as I write this), 2012, of the **Textbook of Obesity.** If we spend a lifetime trying to eat less or not too much and we end up fat and diabetic anyway—as many of us have—it's a good reason to believe that eating too much wasn't the problem and, once again, we're best served looking elsewhere for a solution. This is the beginning of the conversion experience.

Here's how Hafsa Khan, a West Virginia physician, described this troublesome situation to me when I interviewed her in the fall of 2017. All her life, she said, she'd been at best overweight, often obese. She would struggle to lose weight, succeed for a short time, and inevitably regain more. She had her weight under control during medical school but gained twenty-five to thirty pounds during her medical residency. Then she began having children and gained more. After her second son was born, she once again tried to lose her excess weight—upping her gym time and cutting her calories. "I'm eating what I think is healthy," she told me. "Remember, I'm a physician, I'm supposed to know this." When she finally reached out for guidance to a physician friend who is board-certified in obesity medicine, she weighed 235 pounds: "In the last year I've been struggling like hell to lose seven or eight pounds," she told her friend, "when I have seventy to lose."

The journalist Michael Hobbes recounted similar stories in a poignant 2018 **HuffPost** article on the seeming intractability of obesity. The individuals whom Hobbes interviewed were struggling to lose minimal amounts of weight, although in their case remaining clearly and, in several cases, heartbreakingly obese.

"She wakes up, showers and smokes a cigarette to keep her appetite down," Hobbes wrote of one of the women he interviewed. "She drives to her job at a furniture store, she stands in four-inch heels all day, she eats a cup of yogurt alone in her car on her lunch break. After work, lightheaded, her feet throbbing, she counts out three Ritz crackers, eats them at her kitchen counter and writes down the calories in her food journal. Or not. Some days she comes home and goes straight to bed, exhausted and dizzy from hunger, shivering in the Kansas heat. She rouses herself around dinnertime and drinks some orange juice or eats half a granola bar."

This was one of many times this young woman had tried to starve herself to become lean. The last time she tried, a few years earlier, Hobbes wrote, she had kept it up for six months until her mother finally took her to the hospital—still obese, "still wearing plus sizes"—fearful that her daughter had an eating disorder.

The medical orthodoxy accepts this situation as essentially good enough, worth the lifelong effort, by promoting the idea that losing even a little bit of excess weight can bestow "big benefits," which is how the Centers for Disease Control describe it on their website. As little as a 5 percent weight loss—what would be twelve pounds in Hafsa Khan's case—is all that's said to be necessary, by this way of thinking. The big benefits are suppos- edly to our health, as they're clearly not in our girth. Maintaining this little bit of fat loss, so this thinking goes, is surely better than a lifetime of yo-yoing in and out of semistarvation.

Support for this notion comes from the results of a large and influential clinical trial called the Diabetes Prevention Program (DPP). In 2002 the DPP researchers reported that if we take the advice of experts, restrict our calories and control our portions (walk away still hungry from our meals), and exercise for at least 150 minutes a week (say brisk walking or jogging thirty minutes a day, five days a week), we can expect to lose a dozen pounds in a year and maybe maintain an eight-pound weight loss after four years. In so doing, according to the DPP results, we can expect to delay the onset of diabetes by two or three years. We'd have to keep this regime up for a lifetime, or at least until the diabetes sets in and we require medications and eventually insulin to control our blood sugar.*

That's quite a lifelong sacrifice, though, for a payoff that we barely notice and likely won't appreciate when it accrues. If I get diabetes when I'm sixty-five, for instance, instead of sixty-two, I will have no personal awareness of that benefit. It's not as though I would be conscious during those three

---

* The DPP investigators reported this observation as reducing the incidence of diabetes by 58 percent over three years, but the same data can be interpreted as postponing the onset of the disease by several years.

bonus years of health that I had earned my so-far diabetes-free status. That's a lot of work and a lot of sacrifice for little **perceptible** gain. Few who are significantly overweight or obese will consider this kind of benefit worth a lifetime of work (of counting out three Ritz crackers every evening). Promoting the "big benefits" of a 5 percent weight loss is the act of medical and public health authorities who have lost hope. They've lost hope because they're working with naïve and poorly thought-out assumptions about the cause of the disorders— why we're fat, and why we're diabetic or becoming diabetic.*

Among the chief criticisms of fad diets—indeed, among the diagnostic criteria of fad diets—is that they often restrict entire food categories: all animal products, for instance, or all grains, starches, and sugars. This makes them unbalanced, by the

---

*More evidence that this is an old story, repeated again and again, is that Astwood made the same point in the concluding paragraphs of his 1962 presidential address, evoking Brillat-Savarin and **The Physiology of Taste** to do it: "The travail of the obese in trying to slim by diet could not be better expressed than by the statement of a patient to his doctor, recited in 1825," Astwood said: "'Sir I have followed your prescription as if my life depended on it, and I have ascertained that during this month I have lost some three pounds, or a little more. But in order to reach this result, I have been obliged to do such violence to all my tastes and all my habits— in a word, I have suffered so much—that while giving you my best thanks for your kind directions, I renounce any advantages from them and throw myself for the future entirely into the hands of Providence!'" (Astwood 1962.)

conventional thinking, probably unsustainable, and maybe even deadly. (I will discuss this later.) But we can't escape the logic that a successful diet, a diet that **works, must** remove or at least minimize consumption of whatever is causing the ailments or making them worse—specifically, making us fatter and/or more diabetic than ideal, and keeping us that way. What ails us may have no relationship to what we eat, in which case no change in our eating habits is likely to matter. But if it does, we have to identify what it is we eat that's causing or exacerbating the problem and remove it, or at least limit its consumption. If that happens to be a food group and removing it makes the diet unbalanced, so be it. We're clearly better off eating this way as long as what remains in our diet has all the vitamins, minerals, and other micronutrients necessary for health.

Since the mid-twentieth century, dietitians have embraced a belief system in which labeling foods as "bad" ultimately does more harm than good. As a recent BBC article on sugar described it, labeling a food taboo "may only make it more tempting." But what if "bad" foods do exist? Few would argue that labeling cigarettes taboo makes cigarettes more tempting to smoke; no one would argue (or so I hope) that the extreme difficulty of giving up cigarettes or of sustaining nonsmoking status for a lifetime says anything meaningful about the relative benefits of quitting. I can't imagine any rational individual arguing that declaring corn products to be "taboo" for me—labeling them "bad" foods—made me want to eat them more. Even as a child, what I wanted was a life without gastrointestinal distress. If that meant no corn—not even corn on the cob or popcorn at movies, both of which I would have happily eaten to excess—I was willing to accept that reality and pay that price. Before we decide whether labeling a food taboo causes more harm than good, we have to establish whether such foods are indeed harmful, and if so, how that harm manifests itself and why. Only after those questions are answered correctly can we deal with the psychological issues raised by a taboo label.

Whether diets minus entire food groups are sustainable is a slightly more complex question. What seems sustainable is likely to change with time and will be determined in part by the benefits of abstinence. If struggling like hell (counting out Ritz crackers) leads to very little or no apparent benefit—seven or eight pounds lost out of a desired seventy, still wearing plus sizes after half a year of virtual starvation—why struggle? Eating in a way that provides significant weight loss without hunger, though, is likely to be far easier to sustain. If nothing else, the greater benefits are more likely going to be worth the lesser costs. What we are working to sustain is our good health, and if that requires sustaining a particular way of eating, that's what we will work to do. The authorities often criticize fad diets for promising "quick weight loss" in what they think is an unsustainable manner, but these authorities don't understand what it means when a way of eating "works" for those of us who fatten easily.

Don't get me wrong, quick weight loss has its value. "Nothing serves as well as success," as Michael Snyder, a bariatric surgeon in Denver, described this idea to me. But ultimately those who fatten easily, who are predisposed to develop overweight and obesity, want their bodies to work like the bodies of naturally lean people. They'd like to be able to eat to satiety without being fat and or getting fatter. Whether that is too much to ask is another critical question. It may not be possible. But if it is, they'd like to remain relatively lean for a lifetime, without having to consciously, day in, day out, live with hunger, count calories, measure out portions, go to bed hungry, wake up hungry, and deal with the fatigue and irritability that are natural consequences of food deprivation. Sacrifices will be made, but living with hunger cannot be one of them. We can't expect to endure it.

Eating in a way that works does not mean merely losing weight for six months to a year and then regaining it. It's correcting the problem of excess weight, allowing us to eat to satiety without putting on fat or carrying significant amounts of excess fat. If it can do that, it will be sustainable, almost by definition.

When Malcolm Gladwell described the "conversion narrative" of diet book doctors in his 1998 **New Yorker** article on obesity, he notably included losing weight "as if by magic." This is what the diet book author claims to experience in this narrative, and this is what his or her patients supposedly experience as well. Gladwell's article gave the impression that such a narrative was a con, an experience invented merely to sell the book—in short, part of the snake-oil sales pitch. Yet losing weight as if by magic means little more than losing fat, or becoming lean, **without hunger,** relatively without struggle. That it happens quickly is a bonus. That it happens without the inevitable physiological consequences of food deprivation, of starvation or merely semistarvation—i.e., "excessive fatigue, irritability, mental depression and extreme hunger," as Margaret Ohlson, a pioneer in weight-loss diet research and chair of the Food and Nutrition Department at Michigan State University, and her colleagues described it in 1952—is the key.

Such experiences are clearly possible. In **The Importance of Overweight,** Hilde Bruch recounted precisely that of one of her patients, a short, small-boned young woman who was "literally disappearing in mountains of fat." This despite the fact that "everything in [her] life was rated according to whether it would make her fat or help her to lose weight. Going to the beach, bicycle riding, playing golf, or dancing were forced upon her in order to make her slimmer." This young woman described her life as barely worth living. "I actually hated myself," she said to Bruch. "I just could not stand it. I didn't want to look at myself. I hated mirrors. They showed how fat I was."

Under Bruch's guidance, she lost nearly fifty pounds over the course of a summer eating "three large portions of meat" each day with "only some fruits and vegetables in addition." Bruch based the diet on the work of the DuPont Corporation physician Alfred Pennington, who had published his clinical experience with LCHF/ketogenic diets in the medical journals in the late 1940s and early '50s and whose work led eventually to Herman Taller's **Calories Don't Count** and Atkins's **Diet Revolution** and all the LCHF/ketogenic eating regimens that have come since.

"The results were dramatic," Bruch wrote, "not only because her appearance changed but because it gave her the first awareness of some independence from the bite-by-bite supervision she had suffered from up until then. There was also a beginning understanding of her own role in all these difficulties. Until now sentences like 'I don't like it,' or

'I never did it,' had been her final pronouncements that she could not or would not do anything, and this related not only to food but to all other activities. This diet involved having completely unusual meals and she learned, with real amazement, that her taste could change."

Had either Bruch or her young patient chosen to write a diet book pitching meat consumption ("three large portions" daily!) and the absence of sugars, grains, and starchy vegetables as a key to obesity remission, they would have had two choices: (1) use the conversion narrative to describe the benefits and come across, perhaps, as insincere; or (2) tiptoe around what they had actually observed or experienced, despite the fact that their readers would be reading the book hoping to learn how to have precisely such a conversion experience (hoping that what happened to Bruch's young patient would happen to them as well). While Bruch's book was a uniquely thoughtful discussion of many of the issues associated with obesity, it could have been a diet book. Bruch clearly believed that such a meat-rich, carbohydrate-poor diet was a possible solution to obesity and that sugar, starchy carbohydrates, and grains could cause it. "The great progress in dietary control of obesity" since the mid-nineteenth century, she wrote, "was the recognition that meat, 'the strong food,' was not fat producing; but that it was the innocent foodstuffs, such as bread and sweets, which lead to obesity."

At the time Bruch wrote those words, the medical literature was already rife with reports of the remarkable success—what physicians would call the "clinical efficacy"—of diets that restricted these "innocent foodstuffs" and included copious animal products. Physicians working in hospitals and clinics around the world were publishing reports similar to Pennington's: These unbalanced diets restricted in sugars, grains, and starches, fat-rich instead, induced significant weight loss **without hunger.** This was the case in report after report, independent of how many calories the patients in these various institutions were fed, whether fewer than five hundred calories a day (as at the Mayo Clinic) or whether the patients were encouraged to eat as many calories as they could, as was often the prescription. "The absence of complaints of hunger has been remarkable," the Mayo Clinic's Russell Wilder wrote in 1933.

By the early 1950s, physicians at major medical schools were publishing and discussing their variations of these meat-centric, starch-, grain-, and sugar-poor diets for obesity in major medical journals. Often they restricted added fats as well, butters and oils, because they thought that would help people eat less, but they almost always restricted what Bruch had called the innocent foodstuffs. Here's the British endocrinologist Raymond

Greene's version from his seminal 1951 textbook
The Practice of Endocrinology:
Foods to be avoided:

1 Bread, and everything else made with flour
2 Cereals, including breakfast cereals and
milk puddings
3 Potatoes and all other white root vegetables
4 Foods containing much sugar
5 All sweets

You can eat as much as you like of the following foods:

1 Meat, fish, birds
2 All green vegetables
3 Eggs, dried or fresh
4 Cheese
5 Fruit, if unsweetened or sweetened with
saccharin, except bananas and grapes

And here's how Robert Melchionna of Cornell
University's Medical School described the reducing diet that they used at New York Hospital in
Manhattan in the early 1950s: "Concentrated
carbohydrates, such as sugars and breadstuffs, and
fats must be restricted. Diets, therefore, should ex-
clude or minimize the use of rice, bread, potato,
macaroni, pies, cakes, sweet desserts, free sugar,
candy, cream, etc. They should consist of moderate

amounts of meat, fish, fowl, eggs, cheese, coarse grains and skimmed milk." And how about the "general rules" of a successful reducing diet, as published by a physician at Chicago's Children's Memorial Hospital in 1950?

1 Do not use sugar, honey, syrup, jam, jelly or candy.

2 Do not use fruits canned with sugar.

3 Do not use cake, cookies, pie, puddings, ice cream or ices.

4 Do not use foods which have cornstarch or flour added such as gravy or cream sauce.

5 Do not use potatoes (sweet or Irish), macaroni, spaghetti, noodles, dried beans or peas.

6 Do not use fried foods prepared with butter, lard, oil or butter substitutes.

7 Do not use drinks such as Coca-Cola, ginger ale, pop or root beer.

8 Do not use any foods not allowed on the diet and only as much as the diet allows.

In the 1960s, as physicians began holding conferences to discuss the latest developments in obesity research, the conferences invariably included a single talk on dietary therapy. That talk, invariably, would be on the remarkable clinical benefits of LCHF/ketogenic eating. The physicians, psychiatrists, and dietitians at these conferences knew that calorie restriction (eating less) failed, so that was apparently considered unworthy of taking up their time. Not so these diets restricting carbohydrates and allowing significant to unlimited consumption of foods rich in fat and protein.

The most influential of these conferences was held in October 1973 at the National Institutes of Health in Bethesda, Maryland. It was the first conference the NIH ever hosted on obesity. Charlotte Young, a Cornell University professor, gave the only talk on dietary therapy, reviewing the hundred-year history of diets restricting sugar, starchy carbohydrates, and grains, and the results of the multiple clinical trials even back then, including Young's own trials at Cornell. All these LCHF diets, Young said, "gave excellent clinical results as measured by freedom from hunger, allaying of excessive fatigue, satisfactory weight loss, suitability for long-term weight reduction and subsequent weight control."

In short, they worked, as Gladwell might have said, "as if by magic." They resulted not just in weight loss free from hunger, but in weight loss without the other consequences of a body that is being starved for fuel—fatigue or exhaustion. Subjects could eat to satiety, be energized by the experience, and lose weight regardless. Isn't this precisely what we want?

7

A Revolution Unnoticed

Diets that reduce excess fat without hunger
require that insulin be minimized.

Why the magic? What does this experience of weight loss free from obsessive hunger tell us about the makeup of diets that can make this happen, and, perhaps more important, about the relationship between what we eat and why we get fat to begin with? In other words, is it what we eat or how much we eat that's the problem?

Between the mid-1950s and 1970, the answers to these questions were mostly worked out by laboratory researchers studying fat metabolism. They made the critical advances post-1960, following the invention of a laboratory technique (an assay) that allowed these researchers, for the first time ever, to accurately measure the levels of hormones circulating in the bloodstream. The inventors were the physicist Rosalyn Yalow and the physician researcher Solomon Berson. Yalow won the Nobel Prize for the work in 1977. (Berson died in 1972 and so could not share it.) The Nobel committee described Yalow and Berson's assay as bringing about "a revolution in biological and medical research."

It did, but the revolution passed mostly unnoticed by the obesity research community and by those authorities who were advising us on what we must do to achieve and maintain a healthy weight. Not so the fad diet book doctors of the era, but the same authorities were telling us in no uncertain terms that the diet book doctors were quacks. The revelations from this half-century-old research are more important than the latest studies covered in the media that purport to tell us the constituents of healthy eating, and I'll tell you why.

Remember, we're dealing with a disorder of excess fat accumulation, as Bruch and Astwood said, and so we need to understand the physiological processes that regulate fat metabolism in the human body—in particular (borrowing Astwood's phrase) the "complex role the endocrine system plays in the regulation of fat." This then raises the mechanism questions: We know this system is shifted in the direction of storage, and indeed excess storage, but what could explain that shift? And how does this shift in the direction of storage relate to what we eat or how much we eat, such that we can influence it or, ideally, reverse it by diet? The endocrine system does indeed play a complex role in all this, but the answers that are required to successfully treat overweight and obesity by dietary changes turn out to be relatively simple (with the acknowledgment that **relative** is a relative term).

By the 1950s, researchers studying human metabolism (most notably the Nobel laureate Hans Krebs, for whom the famous "Krebs cycle," by which our cells are energized, is named) had already come to understand the basic metabolic systems that work to ensure that the food we eat is made available as a steady and reliable flow of energy to all the cells in our body. The gist of it is that the power plants in the cells (known as mitochondria) that generate the energy we use for life can do so by burning carbohydrates, proteins, or fats as fuel, the three "macronutrients" in our diet.

The endocrine system—hormones and their target enzymes—then plays the critical role in orchestrating what we do with these fuels, when we do it, and for how long. By 1962, when Astwood gave his presidential address to the Endocrine Society, endocrinologists knew that most of the hormones they had identified work to accelerate the release of fat from our fat cells so that cells in the muscles and organs can use it for fuel. These hormones work to make us thinner, in effect, because they work to make our individual fat cells thinner.

Hormones are signaling our bodies to do something—fight, flee, grow, reproduce. It makes sense from an engineering perspective alone that they would also make available the fuel necessary for that action to take place. They liberate fat from our fat cells and prepare the other cells in our bodies to burn that fat for fuel. When you're scared, for instance, your adrenal glands respond by secreting adrenaline into your circulation. That adrenaline not only revs you up to fight or flee, it causes your fat cells to liberate stored fatty acids so that fat is available in the circulation to fuel any fighting or fleeing that might take place. As adrenaline and these hormones linger in the circulation, they keep these fatty acids available just in case. While they do so, they deter the fat cells from taking up fat and storing it. From the perspective of the fat cell, they keep it leaner than it would otherwise be.

The "reverse process, the reincorporation of fat into the depots," as Astwood called it, was found to be dominated by a single hormone. All the other known hormones worked against putting fat into fat cells or back into fat cells; the hormone insulin, as Astwood said, "strongly promoted" it. While physicians and diabetes specialists (even endocrinologists) had come to think about insulin almost exclusively as a hormone that controls blood sugar (most still do), that's like thinking of the conductor of an orchestra as conducting only a single instrument. Insulin does many things in the human body. A primary function is indeed to keep blood sugar under control, but the relevant point for our purposes is that one way it accomplishes that is by also promoting the storage of fat.

Prior to the discovery of insulin in 1921, patients with what we now call type 1 diabetes—the acute form of the disease that typically appears in childhood—would die emaciated and famished, no matter how much food they consumed. But administering insulin to these young patients would bring them back from the brink of death and have them looking healthy again within weeks. It was life-saving. It also appeared quite obviously to be fattening, albeit in a good way. Charles Best, who discovered insulin with his fellow Canadian Frederick Banting, later coauthored a medical textbook that declared this an unmistakable observation: "The fact that insulin increases the formation of fat has been obvious ever since the first emaciated dog or diabetic patient demonstrated a fine pad of adipose tissue, made as a result of treatment with the hormone."

For those who needed more evidence, insulin therapy was also used in the 1920s to fatten up underweight and emaciated patients (those who today we would say suffer from anorexia). It was also used through the mid-twentieth century as a kind of shock therapy in mental hospitals for patients with schizophrenia. These patients typically responded to the therapy by getting fatter—most famously, the Princeton mathematician and future Nobel laureate John Nash, and the author and poet Sylvia Plath. In her fictionalized account of her experience, Plath wrote that she put on twenty pounds with insulin therapy, that she "just grew fatter and fatter." When insulin was administered to patients with the chronic form of diabetes, what used to be called adult-onset diabetes and is now known as type 2 diabetes, they also got fatter. They still do.

While these insights were not embraced by researchers thinking about obesity, their reason was understandable. Yes, insulin clearly seemed to make people fatter in these specific situations, but many or most people diagnosed with diabetes— those with type 2—were already overweight or obese even before insulin therapy. Until the early 1960s and the work of Yalow and Berson, the consensus of opinion among physicians and diabetes specialists was that **all** cases of diabetes had a deficiency of insulin—too little insulin to control blood sugar. This was clearly the case for the acute, childhood form, type 1, so these physicians and researchers assumed it was true for all diabetics. If individuals with diabetes could be obese even if they lacked insulin (necessary to keep their blood sugar in check), it was hard to imagine how insulin played any significant role in having made them, or anyone else, fat.

This is where being able to actually measure hormone levels in the bloodstream made all the difference. Beginning in 1960, in their very first papers using their new insulin assay, Yalow and Berson reported that people who were obese and particularly those who were both obese and diabetic had excessive amounts of insulin circulating in their blood. Not too little, **too much.** Older patients with obesity and diabetes weren't suffering from insulin deficiency; rather, they seemed to be resistant to the insulin they were secreting. This condition is now known as insulin resistance.

Insulin resistance turns out to be fundamental to both obesity and type 2 diabetes—type 2 diabetes more or less **is** insulin resistance—and all the chronic diseases associated with them. When we are insulin resistant, our bodies (the pancreas, specifically) produce more and more insulin trying to achieve the necessary blood sugar control. As this happens, as Yalow and Berson suggested, that insulin will do what insulin does, which is signal fat cells to store fat. The fact that people with obesity and type 2 diabetes are fat is evidence of that. The cells of lean tissue and organs (specifically the liver) might be insulin resistant, even while fat cells remained sensitive to the hormone.

By 1965 Yalow and Berson were describing insulin as "the principal regulator of fat metabolism" and suggesting that the insulin resistance that they were seeing in people with obesity and diabetes might clearly explain why they were fat. When insulin is secreted, it prompts cells throughout the body to take up more blood sugar from the circulation and use it for fuel; it causes liver cells to make fat from glucose and ship that fat out for storage; and it induces fat cells to take up and store any fat for the future. To get fat out of those fat cells, as Yalow and Berson described it, the absolute fundamental requirement was not eating less or exercising more, but lowering the amount of insulin in the circulation. (Eating less and exercising, as I'll discuss, can be inefficient ways of lowering insulin levels.)

To be precise, Yalow and Berson said, getting fat out of our fat cells "requires only the negative stimulus of insulin deficiency." That concept is fundamental to understand. University of Wisconsin researchers studying obesity made a similar declaration in the prestigious **Journal of the American Medical Association:** It could be stated "categorically," they wrote in 1963, that obesity was impossible **without** adequate levels of insulin and that storing **excess** fat "cannot take place" without some insulin around to make it happen and, critically, without the body taking in carbohydrates—glucose—to stimulate that insulin secretion.

In short, by 1965, there were now two competing ideas about how foods and diets can affect our weight and how much fat we store. The conventional wisdom was then and remains still (going back to how the **Textbook of Obesity** phrased it in 2012) that **All diets that result in weight loss do so on one basis and one basis only: They**

reduce total calorie intake. The alternative, the one that's based on biology rather than (supposedly) physics, is: **All diets that result in weight loss do so on one basis and one basis only: They reduce circulating levels of insulin; they create and prolong the negative stimulus of insulin deficiency.**

"I know the math," Roxane Gay says in her memoir **Hunger,** as though that **should** be enough to solve her unruly body and reduce her excess fat. "In order to lose a pound of fat you must burn 3,500 calories." She then goes on to observe that this knowledge has clearly been useless to her.

What I and others are suggesting is that knowing the math is irrelevant. What's necessary to prevent and treat and maybe even reverse obesity is knowing the endocrinology, the hormonal influences and how those in turn can be influenced by what we eat.

8

The Body's Fuel

*When you eat carbohydrates, you raise insulin,
you burn carbohydrates for energy, and you store fat.*

To understand why human bodies accumulate excessive fat, it helps to understand what our bodies are working to accomplish when they are healthy. We are endowed (as are all living organisms) with an exceedingly sophisticated system for surviving and ideally thriving in any contingencies (or at least those we might have faced during the past few million years). This system does innumerable critical jobs simultaneously. The relevant one is that it aims to ensure that all its myriad cells and cell types are properly fueled now and will continue to be adequately fueled in the future, with all the future's attendant unpredictability.

This system has to take the macronutrients (the fuels) available in the foods we eat and those stored in our bodies—protein, fat, and carbohydrates—and maximize their utility. It has to make sure that if the body has too much of one kind of fuel and not enough of others, it makes do and limits harm that might result. Specifically, it has to control our blood sugar after carbohydrate-rich meals because high blood sugar is toxic to cells. The most obvious complications of diabetes—blood vessel, nerve, and kidney damage—are primarily due to the toxic effects of high blood sugar, and they are the reason this disease has to be diagnosed early before irreversible damage is done.

As Yalow and Berson and others were working out the role of insulin and other hormones in fat storage, British biochemists were simultaneously illuminating how our bodies and specifically our cells do this fuel-partitioning job—making fuel available efficiently where and when it's needed—without those hormones. The hormonal system, as I'll discuss, is layered on top to modulate this biochemical system and be prepared for emergencies. As these British biochemists showed, our bodies burn carbohydrates for fuel (specifically glucose, the stuff of blood sugar) when carbohydrates are available, and they burn fat when the carbohydrates have been effectively used up or stored (as a compound called glycogen). This makes eminent sense since our bodies have limited space for storing carbohydrates, about two thousand calories' worth, but they can store relatively huge amounts of fat. Or at least most of us can. The protein is necessary to rebuild and repair cells, and although we don't tend to think of it this way, it too can be stored in large amounts as muscle.

Now imagine eating a typical mixed meal containing all three of the macronutrients—protein, carbohydrates, and fat (leaving aside alcohol for the moment). The carbohydrates break down into glucose and enter the circulation, and your blood sugar (glucose) rises. That glucose has to be used for fuel or stored quickly to minimize the toxicity of this quickly elevating blood sugar. The fat can be stored while that happens and then used for fuel later, and the protein, ideally, will be used for cell and tissue repair.

Insulin is the hormone primarily responsible for orchestrating all this. It prompts cells in your lean tissues and organs to take up carbohydrates and use them for fuel; it inhibits them from burning fat and lets that fat escape back into the circulation, where it can be returned to storage. Insulin simultaneously causes the fat tissue to hold on to fat and the muscle cells to do the same with protein. Protein consumption also stimulates secretion of two other hormones, glucagon and growth hormone, the former of which will work to limit fat storage, while the latter will help promote growth and repair.

As we finish burning off or storing (as glycogen) the carbohydrates we've consumed, as our blood sugar is under control and now coming down, so should insulin. With insulin decreasing, the fat tissue will eventually experience that negative stimulus of insulin deficiency, and the fat cells will release the fat from storage—they will mobilize it—and we will burn that fat for fuel. This is what happens or should happen between meals; it happens overnight while we're sleeping, and it will happen for days, weeks, or even longer if we have to survive a lengthy famine or self-imposed period of fasting. This cycle in which tides of carbs and fat alternatively fuel our cells, moving into and out of storage in the process, became known as the Randle cycle after Sir Philip Randle, the British biochemist who led this work in the 1960s.

Nutritionists and dietitians of the conventional school of thinking have been instructed and will tell us that carbohydrates are the **preferred** fuel for our bodies and our brains, thus implying that they are indispensable. But these nutritionists and dietitians are thinking about it the wrong way. The observable fact is that when carbohydrates are available in our diet, we do use them for fuel and we use them first. Whether or not the body and the brain somehow prefer using carbohydrates for fuel, the fact is that we have little or no choice. Since we have such limited storage space, our bodies have three options: Use the carbohydrates for energy, which at least puts them to use; turn them into fat, which the liver will do if necessary; or dispose of them in our urine, which used to be the diagnostic symptom of diabetes prior to the invention of more sensitive tests that can measure glucose levels directly (or indirectly) in the blood.

Once again it will help to quantify what we're talking about, to establish the actual size of the phenomenon, so we can understand it, specifically why controlling these carbohydrates is so critical and tends to take precedence over other jobs that insulin does, particularly in our modern eating environment. So if you're healthy (i.e., **not** diabetic) and you have **not** just eaten a carbohydrate-rich meal, you have around **a teaspoon's worth** of carbohydrate (glucose) circulating in your blood.* That's what the body considers a benign amount of blood sugar. That's about four or five grams' worth of glucose in your blood or about twenty calories' worth. You'll be diagnosed as diabetic if your blood sugar levels while you're fasting (i.e., in the morning, before breakfast) are even moderately above that level: maybe **a teaspoon and a half** of glucose, or the equivalent of about thirty total calories of glucose circulating throughout your entire body. That very small number is the

---

* The calculation is simple. An average healthy human has about five liters of blood and a healthy blood sugar level, on average, is between 60 and 100 milligrams/deciliter. Multiply five liters times 100 mg/dl, and you get five grams of glucose circulating in the blood during fasting. More, of course, after meals. I am indebted to Allen Rader, a physician and obesity medicine specialist in Boise, Idaho, for pointing this out to me and am a bit embarrassed that I hadn't realized it earlier.

elevated blood sugar that causes so much damage in diabetes and that so many drugs are deployed to control.

If we follow conventional ideas about a healthy diet, we will consume about half our daily calories from carbohydrates, perhaps 1,000 to 1,500 each day, or 50 to 150 times more carbohydrates than are circulating in our bloodstream at any one time. That represents a significant engineering problem for the human body. These carbohydrates will enter the body in waves at mealtimes and from snacks and whatever beverages we're consuming, but they can't be allowed to accumulate in the bloodstream or the consequences will be dire. Yet the storage capacity, as glycogen, is minimal and may be full already. It helps that carbohydrate-rich foods tend to contain significant fiber (or at least this used to be the case before the food industry perfected the art of processing carbohydrates and removing all the fiber, let alone making sugary and other carbohydrate-rich, fiber-free beverages like beer). The fiber will slow down digestion and absorption of the carbohydrates and the time it takes for them to enter the circulation. But once they're in the circulation, they have to be dispensed with quickly.

Our bodies begin to handle this engineering problem by having the pancreas secrete insulin even before we eat. This is known as the cephalic phase insulin release, with **cephalic** meaning "pertaining to the head" or, in this case, what the head and brain are doing rather than the body. The insulin prompts our fat cells to hold on to fat and our lean tissue to take up glucose and burn that for fuel because the body is assuming that more is coming. Just by reading the words **fresh, hot doughnuts,** for instance, you very likely thought about eating, and this cephalic process was put in motion. You may also notice that you're salivating a bit, which is the classic reaction that Pavlov described in dogs—another cephalic phase effect. All these effects prepare the body for the flood of carbohydrates and other macronutrients that it now expects.

The pancreas continues to secrete insulin and the levels in the circulation continue to rise as we start eating, even before the food hits our stomach and we begin to digest and absorb it into our circulation. Once that happens and the tide of blood sugar begins to rise, the glucose stimulates the pancreas to secrete still more insulin. All through this process, the insulin is inducing cells in lean tissue and organs to take up the glucose as quickly as possible, and to store or burn it up for fuel. It's causing those cells to burn glucose rather than fat (fatty acids), and it's stimulating fat cells to take up and continue to hold on to fat.

In essence, our bodies make a calculated decision with each meal we eat. They are maximizing health and utility in the short term with the expectation that the long-term consequences can be minimized. We deal with the immediate problem—this flood of carbohydrate and the damage done to our cells by pumping large amounts of glucose through the mitochondria and the Krebs cycle—in part by putting off until later any problems that might arise from the storage of the relatively benign fat that is consumed with the carbohydrates or made from the carbohydrates. Once the carbohydrate situation is under control, the tide of insulin drops (or it does if you're healthy); the fat cells now see the negative stimulus of insulin deficiency and release fat into the circulation, where the cells of lean tissues and organs can and will take it up and use it for fuel. The same insulin deficiency signal causes cells of lean tissues and organs to also burn the fat for energy.

When this system is working well in lean, healthy individuals, it's highly adaptive. Metabolism researchers refer to it as metabolic flexibility. We shift back and forth easily from burning fat to burning carbohydrates: As the carbohydrates come in, the fat is stored. As the carbohydrates are depleted, the fat is mobilized and takes its place as an energy source.

All that is fine, except that this wonderfully dynamic system is dependent on insulin and the negative stimulus of insulin deficiency to function correctly, and that signal can be disrupted with relative ease by what we eat and how we live in our modern world. Without that negative stimulus of insulin deficiency—if insulin remains elevated above some unknown baseline threshold—we will store fat. Our systems, as Hilde Bruch phrased it, will be shifted in the direction of fat storage and away from oxidation (i.e., burning that fat for energy).

This is a critical problem. Excess fat, specifically above the waist, is an exceedingly good sign of insulin resistance, in which case insulin is indeed elevated higher than it should be and elevated for longer than it should be. Those who are insulin resistant are in fat-storage mode (which is the kind of phrase used by diet book authors but one that is nonetheless biologically appropriate) for much longer in the day than ideal and will be predisposed to hold on to fat rather than mobilize it or burn it. They will fatten easily, at least until their fat cells also become insulin resistant, at which point their weight will plateau. As Yalow and Berson noted, it wouldn't take much insulin resistance for a few extra calories every day to be stored as fat, eventually manifesting itself as obesity. This was clearly an implication. This elevation of insulin, alas, could easily be small enough that it would not be measurable by any assay known to man.

9

Fat vs. Obesity

Pay attention to what textbooks say about why fat cells get fat, not what they say about obesity.

Through the 1960s and into the '70s, understanding human metabolism and fat storage became textbook science, even as the authorities who were telling us how to eat healthy (focused on too much food causing obesity and too much dietary fat causing heart disease) continued to find it of little interest. This understanding has mostly stayed textbook science. Go to your local medical library or college bookstore (or bookshelf, if you're a physician) and find a biochemistry textbook or an endocrinology textbook published after, say, 1980. Look up **fuel metabolism** and **insulin.** In some textbooks, you might have to look under the word **adipocyte**—the technical term for a fat cell—or **adipose tissue.** Then go to the pages specified, and that textbook will explain the hormonal regulation of fuel metabolism, and since fuel storage is part of that process, it should explain what makes our fat cells store fat. It will do so in technical terminology, but the message will be that insulin drives fat storage in the context of the elevated blood sugar that comes with either eating a carb-rich meal or type 2 diabetes.*

Here's the 2017 edition, for instance, of **Lehninger Principles of Biochemistry,** widely considered the most authoritative biochemistry textbook, from the summary of a section on "Hormonal Regulation of Fuel Metabolism":

> High blood glucose elicits the release of insulin, which speeds the uptake of glucose by tissues and favors the storage of fuels as glycogen and triacylglycerols while inhibiting fatty acid mobilization in adipose tissue.

Here's a less technical translation: High blood sugar, which you can have when you either are diabetic or have eaten a carb-rich meal, will prompt your pancreas to secrete insulin, which in turn will

---

* Even biochemistry and endocrinology textbooks tend to drift with the prevailing research fashions. Some simple "truisms" get left behind. In this case, as medical science embraced first molecular biology, then genomics and proteomics and other disciplines made possible by the latest technological innovations, even gut biomics, the study of the bacteria colonizing our GI tracts, textbooks have begun to omit some of this basic science.

prompt you to burn the carbohydrates for fuel, store glucose as glycogen and fat, and prompt your fat cells to store the fat you've eaten and the fat made from glucose and hold on to the fat it already has.

As a reminder of the power of paradigms and dogmatic thinking, the same textbook, on the very same page (939), says, "To a first approximation obesity is the result of taking in more calories in the diet than are expended by the body's fuel-consuming activities." The implication is that our fat cells get fat and fatter because our blood sugar goes up and insulin is elevated, but we get fat and fatter because we eat too much. These are entirely different mechanisms, even though you'd think that we'd get fat and fatter for the same reason our fat cells do. It is, after all, our fat cells that are getting fatter.

I hesitate to use diagrams from human metabolism textbooks in a book meant to be readable for most anyone, but since this is precisely what we want to know, I'm going to do it this one time. We want to know what regulates fat accumulations in fat cells, since, as Bruch noted, when we're overweight or obese, we're dealing with excess fat accumulation in, well, fat cells. Here's how this science looks in a diagram from the 2019 edition of the textbook **Metabolic Regulation in Humans,** written by Keith Frayn of Oxford University (with Rhys Evans). Before Frayn retired, a few years ago,

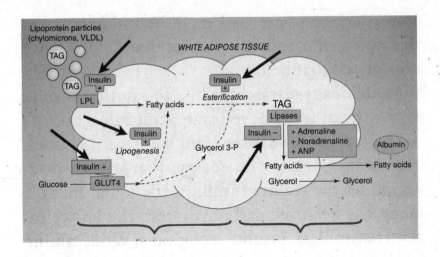

he was considered among the two or three leading authorities in the world on metabolism and particularly fat metabolism.

You can ignore the technical terminology in the diagram and pay attention to the bold arrows that I've added to the figure. As you can see, everywhere the fat tissue is taking up fat, it's insulin that's promoting it—"Insulin +" as it's labeled. When the fat tissue is mobilizing fat, getting fat out of the cells and into the circulation where it can be used for fuel, it's insulin that's inhibiting it ("insulin −") and other hormones (adrenaline, noradrenaline, and ANP in the diagram) that are doing the promoting. (Frayn's **Metabolic Regulation,** too, goes on to blame human obesity on eating too much. The first time I interviewed Frayn, in February 2009, and mentioned that he seemed to have two different mechanisms for excess fat accumulation in fat cells and excess fat accumulation in humans, his immediate response, as I recall it, and I hope I'm not doing him a disservice, was that he had never considered that before.)

Metabolism researchers like to say that insulin is the signal for the "fed state," meaning that it's a signal that we've eaten, and we have fuel available to store and use for energy. That actually oversimplifies the reality: Insulin is the signal that the body has been fed carbohydrates. The fat we eat won't stimulate insulin secretion. (While amino acids from protein are converted into glucose and stimulate insulin secretion indirectly, the protein will also stimulate, as I said, glucagon and growth hormone secretion, so that signal is far more nuanced.) When carbohydrates are consumed and insulin is secreted, it's the carbohydrates that are used for energy, and fat that is put in fat cells. So long as we keep eating carbohydrates and those carbohydrates are absorbed into the circulation, so long as insulin remains elevated and the fat cells remain sensitive to that insulin, it will ensure that fat continues to be stored and to accumulate.

One obvious implication of this basic human physiology is that if we want to get fat out of our fat cells in any biologically efficient way, we have to keep the insulin levels in our circulation low. We have to create that negative stimulus of insulin deficiency, which means not eating carbohydrates. It's all surprisingly simple if we work from the assumption—I would think a very reasonable one—that human physiology, biochemistry, and endocrinology are actually relevant to a problem like obesity and why we get fat. The authorities, for the past half century, have not done that.

What's both fascinating and dismaying about this history is that virtually everyone involved in the diet, weight-control, and health business since the 1960s got at least something important wrong. This was one of the many factors that worked to make a simple message appear to be complicated. Invariably these people made some assumptions based either on their preconceptions about glut-tony and sloth or on the role of dietary fat in heart disease. Some were simply enamored by the physics of thermodynamics and couldn't get away from the idea that what entered the body in excess, whatever that meant, had to be stored as fat. These biases led them to make significant errors in how they interpreted all this evidence.

It didn't help that many of these "experts" had little meaningful scientific training. Typically they were medical doctors who got little more men-toring in doing good science than do plumbers or any other talented artisans. Most of those who had been mentored in science weren't particu-larly good at it. They didn't understand what it meant to be skeptical of their own ideas and so to check and triple-check their assumptions. ("The first principle" of science, as the Nobel laureate physicist Richard Feynman put it so aptly, "is that you must not fool yourself and you're the easiest person to fool.") As a result, these observations about the role of insulin, and the implications that carbohydrates are fattening (specifically, to those who are predisposed to fatten easily), were never taken seriously or considered relevant. They simply didn't fit with the misconceived nutritional notions of the era. When they were taken into account, invariably the researchers interpreted them simplistically and incorrectly.

By 1965, for instance, as low-carb diets were becoming increasingly popular and the science to explain why they worked "as if by magic" had been mostly elucidated, the nutritionists were already saying that the proclamations of physicians advocating for these diets were either "nonsense" (no one can lose weight without eating less) or that the diets themselves were deadly (all that saturated fat!), and that the public dissemination of this dietary guidance would result in "mass murder," as Harvard's Jean Mayer had suggested to **The New York Times** in 1965. Mass murder! Mayer made that statement while clearly understanding the role of insulin in fat accumulation—insulin "favors fat synthesis," he wrote in his 1968 book **Overweight,** while speculating that different levels of insulin and other hormones might have "different effects on the fat content of the body." But Mayer couldn't

leave energy balance behind and convinced himself that those who are fat ultimately get that way
by being physically inactive. The passion for physical fitness that Mayer helped promote began in
the United States in the 1970s and is still going
strong—coincident with ever-higher rates of obesity and diabetes.

The dietitians who were studying and reporting
on the remarkable efficacy of LCHF/ketogenic
eating—weight loss free from hunger—seemed
uninterested in discussing mechanisms that could
explain this remarkable efficacy. If they paid attention to this science, they rarely, if ever, talked
or wrote about it publicly. Researchers who actually studied obesity would later latch on to the
idea that the fat we eat is the fat we store—as
it mostly is—and this, coupled with the notion
being widely promulgated that dietary fat caused
heart disease, led them to advise us to eat less fat
(and replace it with carbohydrates) and that we'd
prevent fat accumulation by doing so. (This might
even work in some people, but at a cost that might
be exceedingly difficult to pay for a lifetime.)
They never made it to the next step in the process, which is that the carbohydrates we eat work
to regulate, through insulin, that fat-storage process and so how much of that dietary fat our fat
cells will store and for how long. One influential
researcher even floated a hypothesis implying that
the body so preferred storing fat to carbohydrates,

thermodynamically, that if a food didn't have fat in it, then it couldn't or wouldn't make us fat. This led to the idea that even sugary beverages—free of fat, as they were—could be consumed to our hearts' content without influencing our waistlines. This was a disastrous misconception, but consumers in this nutrition-obesity-chronic disease world had no protection from bad science and its ubiquitous misapplication.

Even Robert Atkins, who came to fame in this era and knew that insulin was a fattening hormone, still argued in his massively best-selling diet book that his LCHF/ketogenic regimen worked so well because it stimulated some kind of "fat mobilizing hormone," a notion that had been proposed by British researchers in the 1950s and would never pan out. (The reality is that virtually all hormones, with the notable exception of insulin, are technically fat-mobilizing hormones, although they won't mobilize fat when insulin is elevated. The insulin signal overrides that of these other hormones.) When a New York City physician and a Harvard-trained nutritionist joined together to write and publish a scathing critique of Atkins's diet book in 1974 under the imprimatur of the American Medical Association, they pointed out that Atkins's "fat-mobilizing hormone" was a canard and described Atkins's diet as based on "bizarre concepts of nutrition" that clearly shouldn't be promoted to the general public. Then, as an aside to the fat-mobilizing hormone business, they noted that "fat is mobilized when insulin secretion diminishes." That the Atkins diet, an LCHF/ketogenic diet, did among the better jobs imaginable of diminishing insulin secretion was not something the AMA thought should be mentioned.*

* Hilde Bruch got it mostly right, but as I said, she wasn't writing diet books. Here's how she summarized this science in her 1973 book: "Fixation of fatty acids in the adipose tissue for storage depends upon a continuous supply of glucose, and, inasmuch as insulin is required for utilization of this glucose, it is obvious that control of fat metabolism is mediated by glucose and insulin. . . . The implication of this interrelationship is that the excess storage of fat as in obesity, might be associated with, or is the result of, an overproduction of insulin and excessive intake of carbohydrate food, or both." (Bruch 1973.)

10

The Essence of Keto

For those who fatten easily, a way of eating
that restricts an entire food group—an LCHF/ketogenic
dietary pattern—may be necessary and ideal.

Robert Atkins earned his infamy as a diet doctor
in part because his book was selling so well while
promoting the idea of eating large amounts of fat
and, particularly, saturated fats. The establishment
medical doctors and nutritionists may have been
more than a little envious of the former; they were
sincerely worried about the latter. They feared that
Atkins was killing people and conning them in the
process. They didn't trust in the least a concept
that Atkins was the first of the diet book doctors to
fully embrace: ketogenesis and the role of ketone
bodies (less technically, and for our purposes, ketones) and ketosis in a weight-loss diet. This was
what the AMA-sponsored critique was referring

to specifically as a "bizarre" nutritional concept. It was a radical notion then, and it still worries establishment physicians and dietitians.

The idea that avoiding carbohydrate-rich foods was a good strategy if you didn't want to be fat had been around at least since Jean Anthelme Brillat-Savarin in the 1820s. It had become, as I've noted, common thinking. Every woman knew carbohydrates were fattening. Atkins took it one step further and suggested first that the carbohydrates should be replaced by fat (and not just any fat but saturated-fat-rich foods, "lobster with butter sauce, steak with Béarnaise sauce"). He then evoked the concept of ketones and ketosis, what is now called nutritional ketosis—aka keto—as a way to establish whether the diet was actually working, getting fat out of your fat cells and then used for fuel and so out of your body.

Ketones are molecules that are synthesized in liver cells when those cells are burning fat for fuel. They are created from the by-products of this fat burning (oxidation), either from the fat in your diet or from the fat you store when insulin is low enough that the fat is mobilized. Unlike the fat from which they're made, ketones can readily cross the blood-brain barrier, and the brain can and will use them for fuel when carbohydrates are in short supply. That the brain and heart reportedly run more efficiently on ketones than on glucose suggests they may be an ideal fuel for the human body.* Ketosis is what's happening when your liver is synthesizing more than a minimal amount of ketones.

For Atkins, ketones and ketosis were his patent claim, to use Gladwell's phrase, that set his diet apart from the every-woman-knows conventional wisdom. Rather than merely noting that carbohydrates are fattening, and that diets that restrict them but not calories (hence replacing the carbohydrate calories with fat) seem to be a biologically appropriate means to reduce excess weight, he couched his eating plan as a revolutionary diet. As we've discussed, this is a common approach of doctors who write diet books and it has served to complicate simple science, while also offering up a world of speculation that may or may not work to our benefit. In Atkins's case, he wasn't merely claiming his diet fixed a problem by removing a principal cause, which it does; he was providing a unique therapy that could be understood only by reading his book and following his instructions. Those instructions included moving through

---

* Authorities like to say that glucose—blood sugar—is your brain's preferred fuel, but that, again, is because your brain burns glucose for fuel when you're eating a carb-rich diet. It's conceivable that our bodies decided, figuratively speaking, that since our brains use up so much of the energy we generate—around 20 percent—having our brains burn glucose would be necessary to control blood sugar in a high-carb world, even if ketones were somehow a better source, like a higher-octane fuel for your car.

dietary stages, with progressively higher carbohydrate contents if they could be tolerated.

The "induction phase" of the Atkins diet removes all carbohydrates other than those stored as glycogen in meat and the minimal carbohydrates in green vegetables. Most green vegetables fall into the category that nutritionists used to call 5 percent vegetables, which means only 5 percent of their weight comes from carbohydrates that we can digest and the rest is mostly water and some "roughage," which we now call fiber and which we barely digest to make use of for fuel. A cup of broccoli, for instance, has maybe four grams of digestible carbohydrates—sixteen calories' worth—and those carbohydrates are slow to digest and absorb, hence minimizing their effect on blood sugar and insulin. This makes the green vegetables benign from an endocrine perspective, while beneficial from the nutritional point of view. If the green vegetables were eaten along with fatty meats and sauces, that made them more benign still.

Whether Atkins knew it or not, this combination of fatty meat, fats, and green vegetables would come close to most efficiently keeping insulin levels low and prolonging the amount of time that fat would be mobilized from the fat cells and oxidized for fuel, and ketones would be generated. As weight was lost, Atkins counseled, dieters could choose to slowly add back minimal carbohydrates they might miss, so long as their liver continued to churn out ketones—that is, as long as they remained in ketosis.

Atkins called the point at which they stopped generating detectable levels of ketones "the critical carbohydrate level," and the point of his diet was to remain under that threshold. That would be checked by using what are called ketone strips, which can be purchased at pharmacies where they are sold for diabetics, for whom avoiding one particularly severe form of ketosis—known as diabetic ketoacidosis—is critically important to staying alive. Indeed, ketones were first observed in the mid-nineteenth century in the urine of diabetics who were dying from their disease. This is why the medical community has typically seen ketones as signs that something terrible is occurring, as pathological agents, ever since.

It's the wrong interpretation and again overly simplistic thinking, but you can imagine the problem. As I said, establishment physicians still worry about ketones and ketosis, but that's because they don't always read the literature carefully. Steve Phinney of the University of California at Davis and Jeff Volek of Ohio State University are two of the handful of researchers who have actually studied the physiology of ketosis in the laboratory and in clinical trials and so have contributed significantly to our understanding of these molecules and this physiological state. As they have written, ketones are now linked "to a broad-spectrum of health benefits." They are anything but pathological, at least when the body is working correctly.

To understand ketosis and ketogenic diets—keto—you must understand that several conditions must be fulfilled for your liver to synthesize detectable levels of ketones. It has to be burning fat at a high rate, which means insulin levels have to be very low, and that means carbohydrates have to be at least mostly absent from the diet and blood sugar levels have to be at a healthy minimum. One of the many things insulin does is shut down your liver's synthesis of ketones. Again, this makes engineering sense: Insulin in the circulation is a sign that blood sugar is elevated and that cells had better be vigorously metabolizing that glucose, either burning it for energy, storing it as glycogen, or making fat out of it. Ketones, like the fat in our diet, would be neither necessary nor desirable fuel sources during what should ideally be only postmeal periods.

As blood sugar drops, though, and the insulin in the circulation also drops (as it should if you're healthy), fat is mobilized from fat tissue and the liver cells burn that fat. Now the liver's ketone body synthesis goes from "idling in the background," as Phinney and Volek describe it, to generating ketones that can replace glucose as fuel for the brain. Now the body is in "nutritional ketosis." This is a term coined by Phinney to distinguish it clearly from the pathological state of ketoacidosis, when the body lacks all insulin, while perhaps simultaneously distancing it from the aroma of quackery that has always been attached to Atkins for the crime of being a flamboyant (and financially successful) pioneer in this unconventional dietary thinking.

Ketones are measured in units of millimoles per liter, abbreviated as mmol/l. On a typical carbohydrate-rich diet, your ketone level is likely to be about 0.1 mmol/l, which is the product of the liver's ketone body synthesis machinery idling in the background state. If you go twelve hours without eating, which you'll often do in your life— from finishing dinner at seven p.m., say, to having breakfast at a reasonable hour the next morning— your prebreakfast ketone body levels will have tripled, up to 0.3 mmol/l, as your insulin is low and your liver is synthesizing ketones to help feed your brain, if nothing else. Continue to fast for more than several days, and you'll be at 5 to 10 mmol/l.* On an Atkins diet—aka nutritional ketosis—your

---

* Among the key observations in the early 1960s that apparently led Atkins to think of a ketogenic diet as healthy was the fact that our bodies don't distinguish between fuel that is stored and fuel that we've just eaten. Cells will metabolize the protein and fat from either source with no way to know the difference. As such, when our ancestors fasted or had to live through famines, they metabolized primarily or exclusively fat and protein for fuel. The same is true during nutritional ketosis, suggesting that it's a relatively natural or at least benign state, not something to be feared. This has also led Eric Westman at Duke to suggest that rather than say "you are what

ketones might be as high as 2 or 3 mmol/l. After exercise on the same diet, when insulin is very low, you might even hit 5 mmol/l, all relatively low numbers compared to those in diabetic ketoacidosis, the state that so justifiably worries physicians and diabetes specialists.

In diabetic ketoacidosis, fat cells dump their stored fat into the circulation, the liver wildly synthesizes ketones, and carbohydrates are not being taken up and used for fuel at anything like the rate that's necessary. Meanwhile the liver is also generating glucose to use for more fuel. All these fuels are accumulating in the bloodstream, and pathological, metabolic hell is clearly breaking out: Ketone body levels in diabetic ketoacidosis are typically well over 20 mmol/l. This is a condition to be rightly feared, but it is an entirely different physiological state than nutritional ketosis. As I've said repeatedly, physicians and even the expert physicians in this field are prone to overly simplistic thinking, particularly when they are worried that harm will be done.*

For our purpose, ketones and nutritional ketosis

---

you eat," nutritionists and dietitians should counsel patients and clients to "eat what you are"—that is, fat and protein.

* Worth noting is that ketones themselves stimulate some insulin secretion, and the insulin secretion in turn inhibits ketone synthesis. This is a naturally occurring negative feedback loop that prevents ketone levels from getting pathologically high merely from changing our diets.

can be thought of as signs, as biological markers, that fat is being mobilized and burned for fuel rather than stored. Ideally, that would mean you are becoming leaner—the goal, after all, of a weight-loss diet. If the goal is to burn fat without hunger, then nutritional ketosis is a good thing.

Is such a drastic approach wise? That question has driven a half century of controversy over these diets. Surely a way of eating that doesn't restrict an entire food category (and doesn't make your breath smell like acetone from the ketones, which it can) could work as well in reducing excess fat, would be easier to sustain for a lifetime and healthier to boot. Surely that would represent less risk with equal benefit and greater sustainability. Wouldn't it?

The short answer is that it almost assuredly depends on the individual. For the long answer, we have to return to insulin and the statement from the AMA-endorsed denunciation of Atkins: "fat is mobilized when insulin secretion diminishes." This is simple enough, although Yalow and Berson phrased it more precisely when they wrote that the necessary requirement for mobilizing fat from fat cells was "the negative stimulus of insulin deficiency."

Now let's go back to human physiology, metabolism, and endocrinology. As it turns out, this negative-stimulus-of-insulin-deficiency concept comes with two critical caveats. Yalow and Berson were aware of both, but they weren't thinking at the time in terms of the implications for a successful weight-loss diet (let alone writing diet books). The establishment authorities typically paid little attention, perceiving it as not relevant to their gluttony-and-sloth thinking.

First, we all respond to carbohydrates differently. Enormous variation exists from person to person. That's one very good reason why, given the same foods to eat, some of us will grow up to be built like fashion models and some of us will be extremely obese. Moreover, different cells and tissues even in the same individual respond differently to insulin. Here, too, there's enormous variation. When tissues and cells become resistant to insulin, they do so at different rates and different levels of insulin in the circulation. For this reason, as Berson and Yalow cautioned, "it is desirable wherever possible, to distinguish generalized resistance of all tissues from resistance of only individual tissues."

If a physician were to diagnose you as insulin resistant, which is likely if you're fattening easily, that physician would have little if any awareness of how that insulin resistance differed between tissues—whether your fat cells, for instance, continue to respond to insulin even while the other cells in your body had ceased paying attention to the hormone. Whatever is happening with insulin elsewhere in your body, the fact is that as long as your fat cells **do** remain insulin sensitive and insulin is secreted, your fat cells will be storing fat and your body will be accumulating fat. In other words, as Yalow and Berson pointed out, if you're actually getting fatter, your fat cells must be responding to insulin regardless of what is happening elsewhere in your body. Your fat cells must still be insulin sensitive. It seems to be a precondition of the fattening process.

This leads to the second critical caveat, the regrettable one: Fat cells, in particular, tend to be "exquisitely sensitive" to insulin. Some variation on this phrase came to be used commonly by researchers when they described this phenomenon even in their academic articles. I heard it repeatedly in my interviews of those researchers who made the effort to study fat metabolism. It means that fat cells sense and respond to the presence of insulin in the circulation at levels so low that other cells and tissues don't even know it's there, and fat cells continue to respond to insulin long after those other cells and tissues become resistant.

Elevating insulin even slightly above some hypothetical threshold will cause fat cells to enter storage mode. The longer the insulin remains elevated, even if by barely measurable amounts, the longer fat cells will be storing fat, not mobilizing it. For this reason, some of the most prominent diabetes researchers in the world—i.e., the specialists whose purview included paying attention to insulin—had speculated in the 1960s and '70s that having too much insulin circulating in the blood or having fat tissue excessively sensitive to insulin might be the cause of obesity. It might be the reason, as Bruch said, the metabolism shifts too much into storage mode, the reason some of us may trap five or ten or twenty or even one hundred excess calories a day as fat into our fat tissue and others don't. These researchers were just speculating on what seemed like an obvious cause-and-effect possibility, a prime suspect in the mechanism of why we get fat.

In the early 1990s a team of researchers at the University of Texas, San Antonio, methodically measured what being "exquisitely sensitive to insulin" means. In doing so, they identified a threshold level of insulin in the circulation below which fat cells and fat metabolism act entirely differently than they do when insulin is above that threshold. The head of the research team was Ralph DeFronzo, who had pioneered the technology necessary to make this measurement in humans. DeFronzo and his colleagues may have been the only ones in the world capable at the time of making this measurement. As they reported, the "exquisite sensitivity" of fat cells to insulin was their "most striking finding," and (with my apologies for the second and last technical diagram in this book) they presented it in the figure on page 155. This figure may be the single most important figure in the whole diet/weight-loss/obesity discussion. Obesity and fat accumulation, along with their attendant effects on hunger, satiety, and cravings, cannot be understood without understanding the implications of this research.

The figure shows how insulin affects the mobilization of fat (technically, fatty acids) from our fat cells and the use of that fat for fuel at different levels of insulin circulating in the bloodstream. Follow the line on the figure from right to left: from high levels of insulin on the right side to very low levels (the negative stimulus of insulin deficiency) on the left, at the vertical axis. By doing so, what you're seeing is how the fat cells respond to insulin as the concentration of insulin in the circulation drops toward nothing. The horizontal line from over 200 units (μU/ml) of insulin down to about 25 tells us that for most of the range of insulin in our bloodstream, the fat cells remain insulin sensitive and hold on to fat, and the other cells in the body remain averse to using that fat for fuel (oxidizing the fat).

Throughout this range, as DeFronzo and his colleagues wrote in their technical language, insulin inhibits "lipolysis," the breakdown and release of fat from the fat cells and use of fat for fuel. The amount of fat escaping the fat cells and being used for fuel remains the same throughout this range and relatively low. So above a certain level—whether a little above or very high

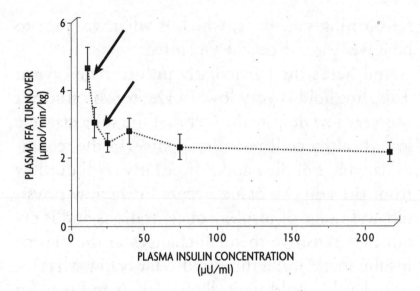

above—the fat tissue remains sensitive to insulin, and the fat within it remains pretty much locked away, trapped. Other cells do not metabolize fat for fuel. Both fat cells and cells in the organs and lean tissues are responding to insulin and acting accordingly.

But then there's the threshold (what the arrows I added are pointing to in the figure). When insulin gets sufficiently low, when the negative stimulus of insulin is sufficiently draconian, everything changes. It's like a switch is thrown. Above the threshold, fat cells hold on to fat. Below it, they release their stored fat into the circulation, and the other cells in the body take it up and use it for fuel. Above the threshold our bodies burn carbohydrates and store fat. Below it, our bodies burn fat. The diet book doctors would say we're fat-burning machines, which is what we want to be if we've got excessive fat stored.

And here's the particularly unfortunate caveat: This threshold is **very** low. In DeFronzo's study, it was very low despite the fact that the subjects were lean, healthy college-age students. (In the technical language of the paper, "[free fatty acid] outflow from the adipose tissue occurs in the low physiological range of insulin concentrations and is exquisitely sensitive to small changes in the plasma insulin level.") The threshold is far below what insulin levels would typically be day in and day out in individuals who are obese or predisposed to get that way—in those who fatten easily. It's exceedingly easy to get above the threshold, which means it's exceedingly difficult to get and stay below it. Those who fatten easily (and who eat sugars, starches, and grains) will spend most of their days and perhaps too many hours of their nights above the threshold, and precious little time—and certainly not enough—below it.

For our purposes, we can think of this threshold of insulin sensitivity in terms of Atkins and his critical carbohydrate threshold. By advising his readers to add back carbohydrates while still checking to see if they were in ketosis, he was essentially telling them to check if their fat tissue, even with the added carbohydrates, remained below this insulin threshold. When we're almost below it, our livers may or may not be synthesizing ketones in large numbers, but we are certainly burning fat for fuel. If we **are** synthesizing considerable ketones and so are in ketosis, we're certainly below the threshold, and the more time we spend below this threshold—day in and day out—the more time we spend burning fat, and the less fat we're storing.

While all of us, lean or fat, **must have** critical carbohydrate thresholds, the more predisposed we are to put on fat, the easier we fatten, the lower the insulin threshold is likely to be or the less time, at least, we spend below it. Atkins's approach of adding back carbohydrates and checking ketones made sense, but it also assumed that this threshold wouldn't change with time, which isn't necessarily the case. It also assumed that maintaining weight loss over a lifetime would be more sustainable and more pleasurable with some carb-rich foods, rather than virtually none.

This may be true for many people and maybe even most. It's probably the reason many people seem to achieve and maintain a healthy weight merely by making sure their carbohydrates are fiber-rich—slow to digest and absorb. This keeps insulin relatively low and, if they're maintaining a healthy weight, below the threshold for these (lucky) people.

But another possibility, all too plausible, is that some of us, at least, will find it easier to eat virtually no carbohydrate-rich foods than to try to eat them in moderation, and that this reality is less about willpower than about, once again, human physiology. Even while some of us may continue to lose weight or maintain a healthy weight while eating "slow" carbs (to use the terminology I first heard from the entrepreneur/author Tim Ferriss), we may find those foods in which the carbohydrates are bound up with fiber and so are digested slowly to be a slippery slope best avoided.

11

Hunger and the Switch

Getting fatter directly influences both hunger and cravings because your brain responds to your body's needs.

We have to live with two realities: that fat cells are exquisitely sensitive to insulin, **and** that this is a threshold effect. The two together have profound consequences for how different foods will affect not just weight but appetites—our hunger and the foods we crave. Those consequences, in turn, speak directly to the question of whether a drastic, supposedly "unbalanced" diet that removes an entire food category may be necessary.

As I suggested earlier, think of this fat-cell, insulin-sensitivity threshold as a switch that's either on or off. When it's on, above the threshold, your fat cells are storing fat; the rest of your body is fueling itself on carbohydrates. When the switch is off, when insulin is below the threshold, your fat cells are mobilizing fat; you're burning fat for fuel; you're getting leaner or at least not getting fatter.

If you're insulin resistant, these dynamics still hold true. But now you have more insulin circulating through your body than is ideal, and the amount of insulin will remain high for longer than ideal. This means you'll spend that much more time above the threshold, with the switch on, storing fat. It's likely this will be the case even long after you've eaten, after blood sugar levels have returned to normal and you might not have carbohydrates (glucose) readily available to burn. Your cells will be primed to burn carbohydrates—that's what the insulin is telling them to do—but blood sugar will already be in the low range of healthy. And while the insulin is instructing the mitochondria in your cells to burn carbs, it's actually pushing those same cells, through the same signaling pathway (as it's technically known), not to burn fat and not to burn protein. Elsewhere, the insulin is causing the fat cells to hold on to fat and the lean cells to hold on to their protein.

In short, when insulin is above the threshold, when the switch is on, your body is running on carbohydrates. They are your fuel. So it makes sense that you'll hunger for carbohydrate-rich foods. This is likely why you may not be able to imagine life worth living without your morning bagel, or your sweets, or your pasta. (For me, it was fresh-squeezed orange juice at breakfast.) Ultimately, as we'll discuss, these carbohydrate-rich foods become your favorites. A likely reason is that your brain has learned to respond to these foods by rewarding you with pleasure when you eat them.

When insulin is below the threshold, when the switch is in the off position, your body is burning the fat you've stored. It will continue to burn fat as long as you remain below the threshold. Now your body has access to plenty of fuel. Twenty pounds of body fat provides fuel for well over two months. Even a lean marathoner like Olympic gold medalist Eliud Kipchoge, who in October 2019 ran the first sub-2-hour marathon ever, at 123 pounds, has enough fat stored to fuel his body on his fat stores alone for a week. Your body is being constantly fed on this supply of stored fat, so it's satisfied. Your appetite will be blunted. The brain has no reason to think more food is necessary. Your body has no need to ingest more food, hence there's little or no urge to do so. You experience weight loss—the burning of your stored body fat—without hunger.

Above the insulin threshold, you have to replenish frequently. You have a limited supply of carbohydrates, and insulin works to keep the carbohydrates you've stored (a maximum of about two thousand calories of glycogen) locked away as well. As your blood sugar drops, you'll get hungry. And because carbohydrates are your fuel above the threshold, you'll hunger for carbohydrate-rich foods.

These dynamics almost assuredly explain the urge to eat between meals, despite how many days' or months' worth of calories we may have stashed away in our fat tissue. It's why we feel hungry when we should, ideally, be happily living off our own fat. It's why we don't feel hungry when insulin is low and we can burn that fat. Another way to think of this is that when you're restricting carbohydrates and insulin is below the threshold, you're not starving your body to get fat out of your fat tissue; you're not at war with your body to lose weight and burn fat, you're working with it, you're allowing your body to do what it will now do naturally.

The relative absence of hunger on these LCHF/ketogenic diets is as consistent an observation as can be found in nutrition science. Remove the carbohydrates and replace the calories with fat, and the stimulus for hunger (and for the obsessive thinking about food that goes with calorie-restricted diets) is lessened significantly. Even those physicians and researchers in the 1960s who were convinced that eating less and semistarvation were the only way to lose weight would often comment in their papers that this didn't mean it wasn't easier to do so on an LCHF/ketogenic diet. As one researcher said in the most famous of the papers from this era, "The satiety value of such diets is superior to diets high in carbohydrate and low in fat." If diets without carbohydrates are more satiating than diets with them, that's just another way of saying that diets with carbohydrates make us hungrier or eat more than diets without. The reason they should is clear.

My favorite example of a physician researcher designing a diet based on this awareness of insulin's role in fat accumulation and the implications for our appetites is James Sidbury, Jr. In the mid-1970s, Sidbury was a pediatrician at Duke University and one of the world's leading authorities on diseases of carbohydrate metabolism—in particular, rare disorders of carbohydrate (glycogen) storage, one of which is named after him. For this reason, it may have been natural for him to think of obesity as a fat-storage disease. Because he was a pediatrician who studied metabolism, the physicians in the Duke medical system would send him their (then) rare cases of children with obesity, hoping he could help them.

Sidbury knew that carbohydrates stimulate insulin and insulin facilitates fat formation and traps fat in fat tissues. He also knew, as he noted in a 1975 book chapter on this work, that kids with obesity crave carbohydrate-rich foods—"crackers, potato chips, french fries, cookies, soft drinks, and the like." Restrict the carbohydrates and feed these kids only fat and protein, he reasoned, and their insulin would come down, and their fat metabolism would work as it does in lean kids. These children would burn their stored fat and lose weight without obsessive hunger and without constantly grazing on carbohydrates. He instructed parents to feed their children with obesity only 300 to 700 calories a day, made up of virtually all protein and fat. The kids lost weight as if by magic. "Many parents do not believe their child can be satisfied with so little food," Sidbury wrote. "Their attitude changes completely," however, when they see the results and, eventually, the "obvious change in the amount of food which satisfies the children."*

Another example of this thinking also dates to the 1970s and comes from George Blackburn and Bruce Bistrian at Harvard Medical School. Bistrian and Blackburn developed what they called a "protein-sparing modified fast" to treat patients with obesity: 650 to 800 calories a day of nothing but lean fish, meat, and fowl. It had effectively no carbohydrates, making it a ketogenic diet, albeit a very low-calorie version. Bistrian and Blackburn prescribed the diet to thousands of patients, as Bistrian told me when I interviewed him in January 2003, and half of them lost at least forty pounds. In one 1985 publication

---

* Regrettably for our understanding of obesity and how to treat and prevent it, shortly after that book chapter was published, Sidbury became director of the National Institute of Child Health and Human Development at NIH. He did not return to his work on the dietary therapy for obesity. In that era, pediatric obesity did not appear to be the critically important subject of research that it clearly is today.

reporting on almost seven hundred patients, the average weight loss was nearly fifty pounds in four months. The patients felt little hunger while on the diet. "They loved it," Bistrian told me. "It was an extraordinarily safe way to get large amounts of weight loss."

But one point that Bistrian made in our interview was critically important: If he and Blackburn had tried to balance these diets by adding, say, vegetables, whole grains, and legumes, meaning the patients would obviously be eating more calories **and more carbohydrates,** the diets would have failed. You'd think more calories would mean less hunger, but it would cause **more.** Bistrian was the first one who pointed out to me the different responses between Ancel Keys's starvation experiment subjects—eating 1,600 calories a day and, well, starving—and the experiences of the patients he and Blackburn were working with, or Sidbury was treating, who were perfectly content consuming far fewer than 1,000 calories a day. "The proof of the pudding," he said to me, "was in the eating."

Alas, Bistrian and Blackburn's thinking, and Sidbury's, was flawed. They were telling their patients—Sidbury's were kids; Bistrian's and Blackburn's were adults—to severely restrict calories because that was what they still thought was necessary. Despite everything they knew about insulin and fat metabolism, they too couldn't escape the trap of energy balance thinking. Because Bistrian and Blackburn were feeding their patients so few calories, this created a problem that appeared to the two researchers to be insurmountable. It had to do with maintaining the weight loss.

For a diet to work for a lifetime, it has to be maintained for a lifetime, and for a diet to work—to get us lean, or relatively so—it has to remove or limit the cause of why we get fat. If the cause is too many calories, then a lifetime of calorie restriction at some level is necessary. If the cause is elevated insulin levels and too many carbs, then a diet that maintains insulin at a low threshold for a lifetime—carb-restricted, high in fat—is necessary. There seems no getting around it.

Bistrian and Blackburn were perfectly aware of this problem. They knew that if their patients went back to eating the way they did, they'd gain the weight back. If they ate more calories but still worked to keep insulin low, they'd be restricting carbohydrates and replacing them with fat. They'd be eating what Bistrian and Blackburn thought of as an Atkins diet. And unless you believed that eating all that fat was benign—as many physicians finally do today—that was unacceptable. Bistrian said this was a primary reason he and Blackburn left the field. They had two choices, they thought: Tell their formerly fat patients to take appetite-suppressing drugs so they could spend a lifetime

battling hunger on a calorie-restricted but balanced diet, or tell them to eat the Atkins/ketogenic
way, to satiety of fat and protein. They considered
neither to be a safe option. "All that saturated fat,"
Bistrian said to me. He and Blackburn turned
their attention to other things. We don't have
that luxury.

The nutritional authorities tend to accept the relative absence of hunger as something that happens
on carbohydrate-restricted diets, but they've shown
typically little interest in trying to understand the
reason. As a result, they have learned nothing from
it. The height of absurdity may have been the 1973
AMA-endorsed critique of Atkins and ketogenic
diets, which went so far as to include "anorexia"—
meaning loss of appetite (not anorexia nervosa, the
chronic condition)—as a negative side effect of
the diet, as something to worry about rather than
embrace. When the authorities did think about
it, they naturally confused the cause and effect to
fit their prejudices. They'd insist that if someone
lost weight eating fat and protein to satiety, then
they must have eaten less, and that was why they
got leaner.

To explain why people would voluntarily accept
semistarvation for weeks or months or years on
end, these authorities would invoke facile rational-
izations that worked to inhibit any curiosity that

might otherwise develop. Among the more common explanations was that an eating pattern without carbohydrates and with plenty of fat was either so boring or so nauseating that people couldn't and wouldn't eat as much as they preferred. My favorite example of this thinking came from Jane Brody, the personal health reporter of **The New York Times,** who was and may still be constitutionally opposed to the idea of LCHF/ketogenic eating. In 2002, in one of her recurring articles attempting to discourage anyone from eating that way, even as an experiment, she explained how it worked this way: "Does it help people lose weight? Of course it does. If you cannot eat bread, bagels, cake, cookies, ice cream, candy, crackers, muffins, sugary soft drinks, pasta, rice, most fruits and many vegetables, you will almost certainly consume fewer calories. Any diet will result in weight loss if it eliminates calories that previously were overconsumed."

Let me offer an alternative, one that is far more likely to be true. When you cut out carbohydrates, you lower insulin sufficiently, mobilize and burn fat, and lose weight. Because you burn your own fat for fuel, your body remains well fed, and you feel no hunger.

One pound of fat, as we discussed, contains about 3,500 calories' worth of energy. If you're losing a pound of body fat a week, easily done when abstaining from carbohydrates, that's 500 calories of fat that you're mobilizing from your fat tissue every day and burning for fuel. That's 500 calories of fat that you would not be mobilizing or using for fuel if you had remained weight stable. You can think of it as though your body were being fed, every day, 500 extra calories of fat. Your cells neither know nor care whether the fuel they're getting comes from your fat stores or from what you just ate for breakfast or lunch or the snack in between. So long as they're being well fed, you shouldn't feel hunger. You're not losing fat because you're eating less. You're eating less, and are content with eating less, because you're losing fat and using that fat to fuel your body.

There is a second, critical implication of this insulin-sensitivity, carbohydrate-tolerance threshold that has to be discussed—it cannot be avoided: If you do anything that boosts your insulin over the threshold, thereby turning on the fat-storage switch, not only will your body return to burning carbohydrates and hoarding fat, you'll hunger for carbohydrates as a result. Cheating by eating a carbohydrate-rich food or meal is very likely to boost you over the threshold and cause the kind of hunger that will tend to keep you cheating. You'll have to work to get back where you were. This is why eating a few french fries won't satisfy your desire for fries and leave you sated. It will very likely create a craving for more, just as an ex-smoker who smokes a single cigarette is less likely to remain an ex-smoker. Physicians who prescribe LCHF/ketogenic eating and who eat that way themselves say it can take days for this hunger for carbohydrates, the urge to continue cheating, to disappear. This is the slippery slope.

There is another complication, courtesy of the implications of the cephalic phase of insulin secretion. Recall that your pancreas will secrete some insulin merely in response to the thought of eating. Just thinking about cheating will have an effect similar to cheating itself, albeit not so dramatic. Unfortunately, the food and beverage industries have devoted themselves to filling our world with stimuli, designed (often by the best advertising minds money can hire) to make us think about eating or drinking carbohydrate-rich foods and beverages. That's the goal of virtually every food- or beverage-related television commercial: stimulate a hunger or a thirst for the product being advertised. Almost invariably that product is carb-rich—pizzas (the crust), fast food (the buns, french fries, and sugary beverages and desserts), beers, soft drinks, and fruit juices. Insulin and fat metabolism are a very likely reason this strategy works. The more predisposed you are to fatten easily, the more exquisitely sensitive to insulin your fat cells are likely to be, the more insulin resistant the rest of your body, and the more profound and self-defeating the effect.

To understand how insulin resistance and cephalic phase insulin secretion might work to undercut our best dietary intentions, imagine two friends walking past a bakery from which is wafting the aroma of freshly baked cinnamon buns. One friend is lean and insulin sensitive. His cephalic phase insulin response to the enticing aroma is minimal. It has little effect on fat mobilization and on whatever mix of fuel—of fat and carbohydrates— his body happens to be burning at the moment. That aroma is going to entice him, but he can walk by the bakery with barely (or almost barely) a second thought.

His friend, though, is insulin resistant, already obese or on his way to getting there. The scent of cinnamon buns will prompt in him a greater insulin response. It will boost him over the threshold and shut down the mobilization of fat and shut down the use of fat for fuel. It will prepare his body to burn carbohydrates. Without having carbohydrates readily available to burn, he will become hungry—perhaps instantaneously—and hunger specifically for carbohydrates. His body's response to the insulin secretion that has been stimulated by his brain is to create a condition of semistarvation (cellular starvation, as Astwood put it) and a response: **Eat!** And the primary, if not only, fuel source his cells will burn when insulin is elevated is carbohydrate.

Driven by a powerful physiological urge to repair

this situation, the insulin-resistant man enters the bakery and buys and eats a cinnamon bun. The lean, insulin-sensitive friend has little conception of what just happened to his pal other than maybe he couldn't resist the aroma of a cinnamon bun. Any lean onlookers would be predisposed to think the fat, insulin-resistant man simply lacked willpower, maybe even moral fortitude, and this would seem to them a very likely explanation for why he was fat.

Willpower, however, had little or nothing to do with it. Those who are lean and insulin sensitive cannot imagine the hunger for carbohydrates that will be induced in those predisposed to fatten, in those who are insulin resistant, when confronted by this aroma and the thought of the bun. It is a subjective experience that lies outside their ability to understand because they never experience it. (Of course, if the lean friend enters the bakery and eats a cinnamon bun as well, no one judges him for it, because, well, he's lean. "Before I got on the plane," as Roxane Gay writes in **Hunger,** "my best friend offered me a bag of potato chips to eat, but I denied myself that. I told her, 'People like me don't get to eat food like that in public,' and it was one of the truest things I've ever said.") For those of us who experience it, though, that hunger is terribly real. We have to understand what causes it if we're to overcome it.

———

What we know about fat metabolism and the insulin threshold implies that many of us who are predisposed to fatten easily must treat this condition like an addiction. Cheating (or thinking about it) begets hunger, which begets more cheating. Eat an apple, let alone a cinnamon bun, and you set in motion a physiological process that creates a hunger for more and a condition in which your body will fatten. The brain prompts the process to begin, then follows along, trained to respond to what the body needs.

This insulin threshold also explains the anecdotal observations that cheating can stall weight loss instantly. In a 1952 research paper, Alfred Pennington of DuPont reported such an anecdote about an obese DuPont executive who effortlessly lost more than fifty pounds on his carbohydrate-restricted diet, eating over three thousand calories a day of meat and green vegetables. He kept the weight off for two years, Pennington wrote, but if he ate **any** carbohydrates, "even an apple," he would begin to fatten again. So for some of us, fixing the metabolic disorder of excess fat accumulation appears to require a total embrace of a different approach to eating, one that restricts an entire food group. It is that simple. Just like smokers who quit cigarettes or drinkers who abstain from alcohol, fixing the condition requires a lifetime of restriction.

The authorities who insist that abstaining from

carbohydrates is an unsustainable lifestyle once again typically do so from the perspective of lean people whose primary fuel happens to be carbohydrates and whose bodies can tolerate carbohydrates without accumulating excess fat. From their perspective, a program that requires living without carbohydrates appears doomed to fail. Why would anyone do it, if another way existed that allowed for the occasional consumption of cinnamon buns and pasta (in moderation, not too much)? But for many of us, there may be no other way. Lean folks aren't like us. They don't get fat when they eat carbohydrates; they may not hunger for them just by thinking about them. They have a choice to live with carbohydrates or not. We don't. Not if we want to be lean and as healthy as we can be.

Like any addiction, this one can be broken, and those who were addicted to another way of eating can learn to be happy and find pleasure in life and, in this case, in eating, without indulging their addiction. Much of what we've learned about making this way of eating work for a lifetime are skills and lessons learned in the addiction world.

When it comes to learning to abstain from carbohydrate-rich foods, we have one advantage that isn't available to those breaking a nicotine or alcohol habit: the beneficial role of fat in these diets. Replacing carbohydrates with fat serves multiple purposes. It keeps calories high and insulin low, which means the body is not semistarving and doesn't respond physiologically as if it were. It also accustoms our bodies to burn fat for fuel. As it does so, hunger and appetites should shift from carbohydrate-rich foods to fat-rich foods. Physiologists have known this happens (at least in animals) since the 1930s, when they reported that rats can be broken of carbohydrate cravings by feeding them high-fat diets. Copious anecdotal evidence implies the same can be true for humans. If your body burns fat for fuel and is used to doing it, fat-rich foods are likely to be what you crave. This may ultimately be the explanation for why butter and bacon are considered, not entirely in jest, to be mainstays of LCHF/ketogenic eating. (It's why, when I do interviews on the radio or on podcasts, I often say I'm one of those people who have convinced themselves that bacon and butter are health foods, and I hope that I am right. In Chapter 12, I will talk in detail about why I believe they are benign.)

Foods that we say we can't live without will change over time. Being both leaner **and healthier** will be sustainable because the food we do consume, unbalanced as they may seem to the experts, will bring us pleasure, and we'll be able to eat them to satiety.

"It is not a paradox to say that animals and humans that become obese gain weight because they are no longer able to lose weight."

That was the assessment of the brilliant French physiologist Jacques Le Magnen in 1984, after several decades of experimentation elucidating the relationship between fat accumulation and hunger, and the critical role that insulin plays in both. Much of our understanding about the role of insulin in preventing us from burning fat and so losing weight is based on Le Magnen's work and thinking. What I've done in this book and my others is merely connect his work on basic physiology to the human diet, to abstinence from carbohydrate-rich foods, and a way of eating that works with your body to reverse fat accumulation, lower insulin, and get fat out of your fat cells.

It should be said, though, that there may be other possible ways of losing fat and maintaining a healthy weight, without having to abstain from carbohydrates: perhaps end-running the system rather than working with it. Rather than fixing the problem by addressing the root cause and fixing that, we can find a hack.

One obvious possibility is to eat diets that are exceptionally low in fat. From the era of Nathan Pritikin in the 1970s to that of Dean Ornish and John McDougall and his starch diet more recently, physicians-turned-diet-book-authors have advocated for very-low-fat diets, primarily for preventing heart disease. These diets allow so few calories from fat—typically less than 10 percent, or a third to a quarter of what most people normally eat—that in practice they require avoidance of virtually all animal foods, which almost invariably come with some fat, even the leanest of chicken breasts. A single serving can boost the daily limit on fat consumption above the recommended maximum in these dietary approaches. Despite the fact that very-low-fat diets replace much of the fat in the diet with carbohydrates and are high-carbohydrate diets, anecdotal observations suggest that some individuals have lost significant weight by following these regimens and have maintained the weight loss. As such, they're worth trying if you buy into the rationale promoted by the authors of these books or if nothing else has worked.

One possibility is that people who lose fat and maintain that loss on these diets are those who manage successfully to starve their bodies of fat. Since the fat we store is primarily the fat we eat, it's conceivable that some individuals can hack this metabolic system by not consuming sufficient fat to supply the fat tissues and the organs (particularly the heart) that normally function on fat. That this fat starvation doesn't eventually create a sustained hunger for fat would be surprising. These diets are even lower in fat content than the diets that Ancel Keys fed his conscientious objectors in his starvation studies, and they may or may not be higher in calories. So eventual feelings of starvation or weight regain may be unavoidable. As with many questions in this weight control field, little meaningful research exists to shed light on this situation. I'm speculating, but starving your body of fat remains a possibility, as the "star McDougallers" who report experiencing dramatic weight loss and serve as anecdotal success stories on John McDougall's website may attest to.

Another possibility is that these diets work because they, too, are restricted in carbohydrates. This restriction is primarily one of quality, not quantity. Even when counseled to avoid fat almost entirely and to live on carbohydrate-rich foods, the individuals following these dietary patterns nevertheless improve the quality of the carbohydrates they consume. They eat carbohydrates that are only minimally processed and that contain considerable fiber, such that both blood sugar and insulin responses are muted. The technical term would be that they eat carbohydrates with a lower glycemic index. (Tim Ferriss memorably labeled them "slow carbs," because we digest and absorb the glucose slowly.) They avoid eating sugar or drinking sugary beverages or carb-rich beverages like beer and milk. They avoid desserts after meals and snack bars between meals. They're doing it to avoid fat, but they're avoiding sugar and refined grains in the process. Hence, it's possible that even on these carb-rich diets, these individuals are improving their insulin sensitivity compared to what it was with their usual diets and are still managing to mobilize fat and get leaner. "We agree people should limit these refined carbs," as Dean Ornish said recently, because of the blood sugar and insulin response. "It's what you replace them with," the type of fats and/or carbohydrates, "that we have a difference."

In this, Ornish is right. Look at virtually any best-selling diet book of the past half century: the very-low-fat (Ornish and Pritikin), the very-high-fat (Atkins), the gluten-phobic (**Wheat Belly** and **Grain Brain**), the lectin-phobic (**The Plant Paradox**), the mostly plant (**In Defense of Food** and **The TB12 Method**), or the almost all plant (**The Starch Solution** and **The China Study**). They all advise, explicitly or implicitly, avoidance of sugars and sugary beverages, and typically highly processed foods of any kind, which means highly processed carbohydrates combined with sugar. They may blame the problems of the modern Western diet on entirely different aspects of the diet—processed foods in general (i.e., Pollan's "foodlike substances"), ultraprocessed foods (a new term), unhealthy fats (however defined), some aspect of the carbohydrate content, the wheat and grains specifically, the fat and oils and salt that are also included in the processing of the carbohydrates, the red meat, all meat, any animal products, and so on—implying that their diets will make us leaner and healthier because we have to give up whichever one of these factors they identify as the cause.

Yet they all agree, whether they state it explicitly or not, that we should avoid highly processed grains and sugar and sugary beverages (and, implicitly, alcoholic beverages like beer), which are the most fattening of the carbohydrates by our understanding of insulin dynamics. Even vegan and vegetarian diet proponents who blame meat and animal products for our eating-related chronic disorders will describe their recommended diet as "healthy" only if it avoids these offending carbohydrates, essentially none of which are animal products.

It's certainly possible that when these diets work, when we eat them and get healthier and leaner, they work because they improve the quality of the carbohydrates we consume and so improve insulin sensitivity, lowering our circulating insulin levels throughout the day and night, and extending the length of time during which we're below the insulin threshold, burning fat rather than storing it. We can think of them as variations on a theme or on a spectrum of carbohydrate restriction, all of which works to lower insulin.

The more mainstream of these dietary approaches are balanced, like the Mediterranean diet. They prohibit highly refined carbohydrates (white flour) and sugars but allow some starches and old-world grains, and they promote beans and legumes (slow carbs). The more extreme diets are those that are ketogenic. As we've discussed, though, the more extreme, more radical the abstinence from virtually all carbohydrates rather than some combination of carbohydrates and fat, the lower the insulin, and the more likely we'll be able to lose weight and maintain a healthy weight without hunger.

Fasting, defined as the voluntary withholding of foods for days or even weeks at a time—intermittent fasting (shorter time periods) and what is now called time-restricted eating (eating all meals in a short six- or seven-hour window of the day)—will extend the duration of time we spend under the insulin threshold, mobilizing fat and burning it for fuel. This will happen regardless of whatever other benefits the fasts might provide. All seem to be effective adjuncts to LCHF/ketogenic eating, and some may be useful on their own for achieving moderate weight loss and improved health (if refined grains and sugars are also restricted). Most of the physicians whom I've interviewed for this book now recommend intermittent fasting or time-restricted eating along with LCHF/ketogenic eating.

I haven't discussed exercise in this book in part because precious little evidence exists to suggest that we can lose any meaningful amount of fat and keep it off merely by increasing the amount of energy we expend through exercise or physical activity. We all may know people, though, who swear they lost weight merely by upping their workouts or returning to them after a lengthy absence. If that's true, then the physical activity had to increase the length of time their insulin levels stayed under the threshold for mobilizing fat.

One way it could have done that is by increasing insulin sensitivity in muscle cells. As this was explained to me by the late John Holloszy, a legendary exercise physiologist at Washington University in St. Louis,* endurance or aerobic exercise will improve insulin sensitivity because the exercise depletes glycogen stores in the muscles, and the cells then essentially work to fill them back up again. By this thinking, the body appears insulin sensitive because the cells are working more vigorously to take up carbohydrates just as they do when insulin is elevated. This effect will last for a day or two after a hard workout, Holloszy explained, or until we've eaten our first carbohydrate-rich meal and hastened the process to conclusion. (If your workout habits are like mine were and the first carbohydrate-rich meal postworkout is a thirty-two-ounce bottle of Gatorade, any benefit to insulin sensitivity will be exceedingly short-lived.)

---

* Holloszy, who passed away in 2018, began studying the relevant metabolism in the early 1960s.

So exercise, too, may help to keep insulin low and fat mobilization high, working to counteract the carbohydrates consumed, but for only a short period of time. This suggests that a more efficient plan would be to abstain from the carbohydrates to begin with.

12

The Path Well Traveled

Given a choice between a hypothesis and
an experience, go with the experience.

Someone asked me the other day how I
was losing weight. I told them I eat less
than 20g of carbs a day. They proceeded to
freak the heck out. Told me how dangerous
it was. (No.) Asked me if my doc knew.
(Yes.) Told me that carbs were essential
to human survival. . . . Finally I was like,
dude, do you really believe I was healthier
90 pounds heavier than I am now? I really
think he wanted to say yes but was worried
that I was going to punch his lights out. He
probably would've been right.

—RACHELLE PLOETZ,
on her Instagram account
#eatbaconloseweight

The question Rachelle Ploetz asked speaks to the very heart of this endlessly controversial subject: "Dude, do you really believe I was healthier 90 pounds heavier than I am now?" Ultimately the goal is to be healthy. Whether ninety pounds are lost or ten, it's quite possible that a way of eating that induces fat loss becomes harmful as the years go by.

Rachelle's experience presents a good case study. Rachelle had wrestled with her weight throughout her life and had tried to eat healthy by the conventional definition. When she began her LCHF/ketogenic program, she weighed 380 pounds. She would eventually lose 150 pounds, documenting it all on her Instagram account and settling in at 230 pounds. Her husband lost seventy-five pounds eating as Rachelle did. Her teenage daughter dropped fifty pounds. They came to believe, as do I, that if they now changed how they were eating, if they went back to eating even "healthy" carbohydrates—say, from whole grains or from beans and legumes (and, of course, cut back on the butter and bacon)—they would eventually gain the weight back. They consider this to be a way of eating for life, out of necessity. Are they healthier for doing so?

When I first wrote about (and still barely understood) the paradox presented by LCHF/ketogenic eating to the medical community in my **New York Times Magazine** cover story in July 2002,

I admitted to trying the Atkins diet as an experiment and effortlessly losing twenty-five pounds by doing so. Those were twenty-five pounds I had essentially been trying to lose every day of my life since I'd hit my thirties, despite an addiction to exercise and the better part of a decade—the 1990s—of low-fat, mostly plant, "healthy" eating. I avoided avocados and peanut butter because they were high in fat, and I thought of red meat, particularly a steak or bacon, as an agent of premature death. I ate only the whites of eggs. Having failed to make noticeable headway, I had come to accept those excess pounds as an inescapable fact of my life. When I changed how I ate—and not, as far as I could tell, how much—those pounds disappeared.

At the time I was simply fascinated by the experience, feeling as though a switch had been flipped (which I now understand to be the case). But I also acknowledged in the article something that remained true for years afterward: my anxiety. Every morning when I sat down to my breakfast of eggs—with the yolks—and sausage or bacon, I wondered whether, how, and when it was going to kill me. I didn't worry about any lack of green vegetables in my diet because I was eating more of them than ever. I worried about the fat and the red and processed meat. Despite all my reporting and my journalistic skepticism, my thoughts on the nature of a healthy diet were a product of the

nutritional belief system that had become firmly
ensconced as I was becoming an adult, the theories
or, technically, hypotheses of what constituted a
healthy diet. Bacon, sausage, eggs (yolks, anyway),
red meat, and copious butter were not included.

"After 20 years steeped in a low-fat paradigm," I
wrote in that 2002 article,

I find it hard to see the nutritional world any
other way. I have learned that low-fat diets fail
in clinical trials and in real life, and they certainly have failed in my life. I have read the
papers suggesting that 20 years of low-fat recommendations have not managed to lower the
incidence of heart disease in this country, and
may have led instead to the steep increase in
obesity and Type 2 diabetes. I have interviewed
researchers whose computer models have calculated that cutting back on the saturated fats
in my diet to the levels recommended by the
American Heart Association would not add
more than a few months to my life, if that. I
have even lost considerable weight with relative ease by giving up carbohydrates on my test
diet, and yet I can look down at my eggs and
sausage and still imagine the imminent onset
of heart disease and obesity, the latter assuredly
to be caused by some bizarre rebound phenomena the likes of which science has not yet
begun to describe.

Little meaningful evidence existed then, as I also noted, to ease these anxieties. A critical fact in this debate, indeed, the reason it continues to exist at all, is that we still have precious little evidence. What we want to know, after all, is whether LCHF/ketogenic eating—rather than, say, a Mediterranean diet or a very-low-fat diet or a vegetarian diet—will not only lead to more or less weight loss but will kill us prematurely.

To establish this knowledge in any reliable manner, we have to do experiments, the finest of which known to medicine are randomized controlled trials. In concept, they're simple: Choose two groups of people at random; have one group eat one diet and the other group eat another diet; see what happens. Which group of randomly chosen individuals lives longer, and which has more or less disease? The catch is that it takes decades for these chronic diseases to establish themselves, and to find out how long we live, and the differences between groups in what is technically known as morbidity (sickness) and mortality (age at death) may be subtle. For these reasons the kinds of experiments that shed light on this question of which are the healthiest eating patterns (for all or some subset of the population) require at least a few tens of thousands of subjects, and then they have to proceed for long enough—perhaps decades—to reliably determine if the subjects are getting more or less heart disease, dying sooner or later, in a way that's clearly the result of what they're eating.

Medicine is a science, so the concept of hypothesis and test still holds, and these clinical trials are the tests of the relevant hypotheses about diet and health. To do these trials correctly, though, would cost a huge amount of money. Many such trials would have to be done, some just to see if the others got it right, and they are almost unimaginably challenging. The concept is simple, the reality anything but. They can fail in so many different ways that some prominent public health authorities have recently taken to arguing that they **shouldn't** be done. They argue that we should trust what they **think** they know about the nature of a healthy diet, and that this knowledge should apply to all of us, whether we are predisposed to get fat on such a diet or not. I respectfully disagree.

Absent this kind of reliable evidence, we can speculate on whether a diet is likely to kill us prematurely or is healthier than some other way of eating (i.e., we'll live longer and stay healthy longer) by applying certain rules, but we must always acknowledge that we are guessing. For instance, eating foods that humans have been eating for thousands or hundreds of thousands of years, and in the form in which these foods were originally eaten, is likely to have fewer risks and so to be more benign than eating foods that are relatively new to human diets or processed in a way that is relatively new. This argument was made famously in the context of guidelines for public health by the British epidemiologist Geoffrey Rose in 1981. If the goal is to prevent disease, Rose observed, which is what public health guidelines and recommendations are intended to do, then the only acceptable measures of prevention are those that remove what Rose called "unnatural factors" and restore "'biological normality'—that is . . . the conditions to which presumably we are genetically adapted."

Remove and **unnatural** are the operative words. Removing something unnatural implies that we're getting rid of something that is likely to be harmful. Take, for example, the advice that we shouldn't smoke cigarettes. We have very little reason to think that removing cigarettes from our lives will do physical harm, because there's nothing "natural" about smoking cigarettes. They're a relatively new addition to the human experience.

If we're adding something that is new to our diets, hence "unnatural," thinking it will make us healthier, we're guessing that the benefits outweigh the harms. There are likely to be both. Now we have to treat that new thing just as we would a drug that we think is good for us and that we're supposed to take for life (say, a drug that lowers our cholesterol levels or our blood pressure). How do we know it's safe, even if it seems to be beneficial in the short term?

All this is a judgment call and depends on

perspective. One reason all diet authorities now agree more or less that we should cut back on our consumption of highly processed grains (white flour) and sugars (sucrose and high-fructose syrups) is that these refined grains and sweet refined sugars are relatively new to human diets. We assume that no harm can come from **not** eating them and perhaps quite a bit of good. Eating or drinking sugar, for instance, might have benefits in the short run—the rush of energy might fuel athletic performance or allow us to perform better on a test in school—but that doesn't tell us whether the long-term consumption is to our detriment. Health authorities have mostly come to believe it is.

The idea that we should all eat tubers, like sweet potatoes, as proponents of the paleo diet suggest, is based on the assumption that our hunter-gatherer ancestors ate them for a couple of million years, implying that they are safe. Some paleo advocates take this assumption a step further and propose that we'd be healthier eating tubers than not. But they're only guessing. It may be true, or maybe it's true for some of us but not for others. We have no way to tell, short of doing one of those incredibly expensive, unimaginably challenging clinical trials.

When we're told that we should consume more omega-3 fatty acids (a kind of polyunsaturated fat in fish oil and flaxseeds, among other sources) and fewer omega-6s (another kind of fat), it is based on the assumption that this shift in the balance of fats we ingest will make us healthier and live longer. In this case, researchers have done a few long-term trials to test the assumption, and the results have been mixed: Maybe they do, maybe they don't. Nonetheless, we continue to hear that we should eat more omega-3s and fewer omega-6s because we currently consume a lot of omega-6s in our diets (from corn and soybean oil, conspicuously, and from eating animals that have been raised on corn and soybeans), and that's considered unnatural. By this thinking, we are not genetically adapted to have such a high percentage of fats from omega-6s. It might be the correct assumption, but we don't know.

One reason I and others promote the idea that eating saturated fat from animal products is most likely benign is that we've been consuming these fats as a species for as long as humans have been a species. The evidence isn't compelling enough to convince **us** that this assumption is likely to be wrong. We may or may not have been consuming as much of these saturated fats, but we can presume we are genetically adapted to eating them. They are "vintage fats," to use a term I first saw employed by Jennifer Calihan and Adele Hite, a registered nurse, in their book **Dinner Plans: Easy Vintage Meals,** and they include some vegetable oils—from olives, peanuts, sesame, avocado, and coconuts—and all animal fats in this category. Calihan and Hite contrast them to "modern fats"—margarine; shortenings of any kind; and industrially processed oils from rapeseeds (canola oil), corn, soy, cottonseed, grapeseed, and safflower. Vintage fats, by this thinking, can be trusted to be benign. Modern fats, not so much.

This is also why we believe that meat from grass-fed, pasture-raised animals is healthier for us than that from grain-fed, factory-farmed animals: The fat content of this meat will be more closely aligned to that of the animals our ancestors ate for the past million or so years. It will be more natural. (Perhaps more important, it is a way of eating that does not support the cruel and inhumane treatment that is common to factory-farming operations.) New foods or old foods in unnatural forms are more likely to be harmful than those foods to which we are presumably genetically adapted.

This belief also, ultimately, underpins the conventional thinking that a healthy diet includes ancient grains—quinoa, for instance, or couscous—or brown rice and whole grains rather than highly refined grains like white rice and white flour. Even without knowing any mechanisms for why this might be true—gluten content or glycemic index (how quickly or slowly the glucose hits our bloodstream)—and absent, once again, any meaningful experimental evidence, the assumption is that our ancestors ate these grains for maybe a few thousand years, in the form in which we're eating them. Hence they are likely to be benign, at least for people who are predisposed to be lean and can tolerate a higher carbohydrate content in their diet.

The caveat, of course, is that definitions of **natural** and **unnatural** can depend on the perspective of the nutrition authority. When we're parsing the latest diet advice, we have to make judgments about how the proponents of the advice define **natural** and **unnatural.** Are ancient grains natural because some populations (but not all) have consumed them for thousands of years, more or less since the invention of agriculture? Or are all grains unnatural because we've been consuming them for only a few thousand years, since the invention of agriculture? Are we safe adding something presumably natural to the diet (ancient grains or tubers or omega-3 fatty acids), or is it a better idea to remove only the unnatural elements (refined grains, sugars, some of the omega-6 fatty acids)? I think the latter is the safer bet. But this, too, gets complicated because as we remove sources of energy from the diet, we have to replace them.

What might be the most complicating factor in how we think about how we eat is the influence of the latest news, the latest media report on the latest study that is making a claim sufficiently interesting to constitute news. By definition that is what's **new,** which means it either adds significantly to the conventional wisdom or contradicts it or speaks to whatever diets have indeed become particularly faddish these days.*

The best reason to ignore the latest study results, the latest media reports suggesting we should eat this and not that, is that the interpretation of these latest studies is most likely wrong. A discussion highlighted in the media these days is what science journalists refer to as the "reproducibility crisis"—some large proportion of the studies that are published either get the wrong results or are interpreted incorrectly or maybe both. If we include those studies that are just meaningless, only one in ten or one in twenty studies (that make the press or appear on your home page) may be worth our notice. This percentage may be even smaller in nutrition and lifestyle research, in which the researchers are so poorly trained and the research so challenging to do. This is one reason the committees that decide on Nobel Prizes traditionally wait decades before acknowledging work to be prizeworthy. Far more often than not, if we wait long enough, we'll see other studies being published

* After my 2002 article suggesting that Atkins was right all along, I was accused of taking a contrarian perspective not because I really thought the evidence supported it, but because it was more newsworthy and would earn me a large book contract. Reporting that the conventional wisdom was indeed right would not. The editors of **The New York Times Magazine** might not have even published such a version because it wouldn't have been news.

making the opposite claims of whatever we're reading today. We won't know which is right until long after their publication. Perhaps never.

"Trying to determine what is going on in the world by reading newspapers," as a famously clever screenwriter/director/journalist named Ben Hecht once wrote, "is like trying to tell the time by watching the second hand of a clock." The same is true of research and science. Trying to tell what's true by looking at the latest articles published in a journal—and particularly in nutrition—is another fool's game. The best idea is to attend little to the latest research and focus instead on the long-term trends, the accumulation of studies (one hopes, interpreted without bias), even if the long-term trends rarely, if ever, appear in the news.

Since the heyday of the Atkins diet in the 1970s, authorities have refused to accept the notion that LCHF/ketogenic eating is safe. (And those that do promptly lose their standing as an authority by doing so.) They believe that the fat content in the foods we think we should eat instead of refined carbohydrates and sugar is too high, so arguably unnatural. Those of us who promote this way of eating can speculate that many hunter-gatherer populations lived on vaguely similar diets and perhaps even in a state of ketosis—the Inuit, pastoralists like the Maasai warriors in Kenya, Native

Americans in the winter months—but we're just speculating. The unusual aspect of these diets leads to legitimate questions about risks outweighing the benefits. This is as it should be.

No matter how much weight people might lose, no matter how easily, the orthodox medical opinion remains that these diets will kill us prematurely. Generations of physicians, medical researchers, dietitians, and nutritionists have been taught to believe (as was I and probably you, too) that we know what a healthy diet is. We know it because this is what healthy people tend to eat. They eat fruits, vegetables, whole grains, pulses (legumes), such as lentils, peas, and beans—mostly plants and plenty of carbohydrates. They avoid red meat and processed meats, and the fats they eat tend to be unsaturated, from plant sources rather than animal. Any radical deviation from this way of eating, regardless of weight loss, according to the consensus of medical opinion, is likely to be unsustainable and ultimately to our detriment.

This is the reason the authorities convened annually by **U.S. News & World Report** to judge diets and tell us what to eat rank LCHF/ketogenic diets as among the least healthy imaginable, regardless of the copious research and clinical experience that now argues quite the opposite. This is why two of the more media-savvy proponents of conventionally healthy eating—David Katz, a physician formerly associated with Yale University, and the former **New York Times** columnist Mark Bittman, a best-selling cookbook author—thought it appropriate to suggest recently in **New York** magazine that losing weight on LCHF/ketogenic eating (let alone maintaining weight for a lifetime) was analogous to getting cholera, an often fatal, infectious diarrheal disease. "Not everything that causes weight loss or apparent metabolic improvement in the short term is a good idea," they wrote. "Cholera, for instance, causes weight, blood sugar, and blood lipids to come down—that doesn't mean you want it!"

Despite the hyperbolic rhetoric, Katz and Bittman have our best interests in mind. Their concern is a legitimate one. The world is full of things we can do or take—medications and performance-enhancing substances—that will reverse and maybe even correct some symptoms of ill health in the short run, but will shorten our lives or ruin them if we take them for years or decades. The first rule of medicine is not actually to do well by your patient, but to do no harm. That's the Hippocratic Oath. As a recent **New York Times** op-ed said about a drug that seems to do a remarkable job of quickly easing serious suicidal depression, "questions also remain about the safety of long-term use."

Questions will always remain about the safety of long-term use . . . of anything. Imagine that you decide to take up running as a hedge against aging. Whether you think about it this way or not, you are implicitly making a judgment about the risks and benefits of the endeavor. Would you suffer more or less damage to your joints, for instance? Will you live longer by stressing your system in these workouts, or will they kill you prematurely? Marathoners die of heart attacks, too, occasionally young. Jim Fixx, author of **The Complete Book of Running,** a best seller in 1977, tragically died of a heart attack while out for a run. He was fifty-two years old. The conventional wisdom is that there are few things we can do that would be better for us, but we'll never know for sure. We know that endurance runners seem to be very healthy, but that may not apply to us.

An almost universal misconception about nutrition and modern medicine—one shared by authority figures, physicians, and the journalists who cover the field—involves when clinical trials are necessary to guide our decisions and when they're not. You do not need a clinical trial (costing tens of millions of dollars with tens of thousands of subjects) to tell you whether LCHF/ketogenic eating, or any regimen from vegan to carnivore, will allow you to achieve significant weight loss easily, without hunger, and make you feel healthier than you did. You can try any of these diets yourself and find out. It doesn't matter what clinical trials conclude. What matters is what happens to you. Try changing the way you eat, and you will find out, just as you can take a new prescription drug and learn relatively quickly whether it helps what-ever ails you and makes you feel better. Clinical trials are necessary to tell us about the long-term risks and benefits of one way of eating versus another—vegan, say, compared to carnivore, the two extremes—not the short-term. Those we can learn about reliably on our own.

"Is it safe?" is always one of two ultimate ques-tions when considering a change of diet or lifestyle, particularly with the goal of preventing chronic disease. "Does it fix what ails us?" is the other. The two questions are so intimately related that we cannot discuss one without the other.

This is one of the many conspicuous problems with the argument that LCHF/ketogenic eat-ing is simply too risky, if not for the short term, then for the long. The authorities who make this argument assume, as we've discussed, that we have viable alternatives, that we can achieve and maintain a healthy weight via any number of di-etary approaches (so long as we eat less), like the Mediterranean diet, which they assume to be safe. For them, the observation that lean, healthy peo-ple eat this way—not all of them, though—seals the deal. To believe that it applies to all of us, you have to believe that those of us who fatten easily, as I've discussed and disagreed with strongly, are no different from those lean folks physiologically and hormonally.

By this orthodox thinking, LCHF/ketogenic

eating is just another of many routes to doing what's necessary: restricting calories and eating less. It's seen as a particularly radical way to accomplish that, and radical ways to do anything are unnatural and entail, by definition, considerable risk, hence a relatively high likelihood of doing harm. According to orthodox thinking, eating a conventionally healthy diet as lean and healthy people appear to do, but less of it, is clearly an alternative for heavy people, one they can assume to be safe. These authorities simply will not confront the possibility that eating less or not too much of a conventionally healthy diet will not fix what ails many of us. If eating a conventionally healthy diet but less of it, and achieving and maintaining a healthy weight by doing so, is not a viable reality **for us,** then this argument falls apart.

It's also critically important to understand the basis of the faith upholding these arguments. The authorities who make them—whether they are the experts convened for **U.S. News & World Report** or the U.S. Department of Agriculture's dietary guidelines or the Katzes and Bittmans of the world or the well-meaning friends ("dude!") who advise us to ease off the bacon—derive their opinions not from experience but from theoretical concepts about a healthy diet. They have merely embraced, as virtually all of us once did, the conventional hypotheses about the nature of a healthy diet. This way of thinking seems intuitively obvious and

seems to work for **them.** In this sense, it's helpful to think of the half-century-old controversy about the nature of a healthy diet as a conflict between hypothesis and experience.*

On the one hand, we have ideas about how best to eat to be healthy, ideas we think are true or that seem to be true. On the other, we have what physicians observe in their clinics and what happens to us, what we experience, when we try different diets. The conventional wisdom on nutrition is dominated by the hypothesis that saturated fats cause heart attacks by raising cholesterol levels, specifically the "bad cholesterol" in low-density lipoproteins (LDLs). This hypothesis has dominated orthodox thinking on diet and health, much as the one ring in J. R. R. Tolkien's **The Lord of the Rings** "rules them all." Hence, eating polyunsaturated fats from corn, soy, or canola oil, instead of saturated fats, by implication, will make us live longer. The ideas that we should avoid animal products (red meat, eggs, and dairy in particular), that they do us harm, and that we will live longer and healthier lives if we eat a mostly or all-plant diet are also based largely on the fear of saturated fats.

Physicians and dietitians are expected to base

---

* I owe this way of thinking about the diet-health conflict to Martin Andreae, a physician in British Columbia, who made this observation when I interviewed him in the fall of 2017.

their diet and lifestyle advice on these hypotheses, but they have no way to know whether their advice makes a difference. When a patient dies, as all eventually will, regardless of age or cause of death, regardless of whether her cholesterol levels changed or not, the physician is privy to no information about what role the low-fat diet might have played. By the same token, should I die tomorrow or in my hundredth year, my next of kin will not know if my unconventional high-fat eating shortened my life or lengthened it. (Critics of my nutrition work will insist that the fat killed me prematurely, regardless, but they'll be guessing.) Maybe Jim Fixx would have had his tragic fatal heart attack a decade younger had he not taken up running. Maybe he would have died in his early fifties regardless. We'll never know.

Even if we had strong clinical trial evidence to support these hypotheses, which we don't, we wouldn't know the answer to these questions. The hypotheses and the evidence on which the authorities come to these conclusions—i.e., to embrace these assumptions—suggest only that we're more likely to live longer if we eat conventionally healthy diets and exercise, not that we will. So we will have to make a risk-benefit analysis as to whether the likelihood that we'll live longer makes it worth engaging in the relevant behavior **for the rest of our lives.** This raises another obvious question: If the authorities are right, for instance, that eating saturated fat will shorten our lives, can we quantify it? How much longer can we expect to live if we restrict our fat consumption?

This is yet another question the authorities seem to avoid, perhaps because the answer is not to their liking. If the conventional wisdom is right and eating saturated fat raises your LDL cholesterol (as for many of us it will) and so gives you a heart attack and kills you prematurely, how many years of life would you have gained if you avoided fat-rich foods and particularly those with saturated fat, or replaced at least some of that saturated fat (from animals) with polyunsaturated fats from seed oils, as the authorities concerned with our heart health advise? In other words, assuming the experts are right, what kind of culinary sacrifice is our fear of saturated fat worth?

As I noted in my 2002 **New York Times Magazine** article, the answer to that question was worked out long ago by three groups of researchers, all in agreement: at Harvard (published in 1987), at McGill University in Montreal, and at the University of California San Francisco (both 1994). These researchers estimated the benefit to longevity if we cut our fat consumption by a quarter and our saturated fat consumption by a third from what we might have typically eaten back then, lowering our cholesterol significantly, and they all concluded that absent other serious risk factors for heart disease, we'd live on average from a few days to a few months longer.

As one of these researchers pointed out to me when I interviewed him, the added time is not in the prime of our lives but rather at the very end of our lives. This seems obvious, but it's a point worth pondering. Instead of dying, say, in March of our seventy-fifth year, we die in April or May. A ninety-year-old is likely to get a few more months being ninety or maybe will make it to ninety-one. That could be a good thing when you're ninety, or maybe not, depending on the quality of your life at the time. A sixty-year-old is likely to gain only a couple of extra weeks. It's not even clear that this dietary intervention prevents any heart attacks. Even in the best of all worlds, it may delay them merely by those few weeks or months.

After the 1987 Harvard analysis was published in the **Annals of Internal Medicine,** Marshall Becker, a professor of public health at the University of Michigan, suggested that avoiding fat or saturated fat to prevent heart disease is "analogous to stewards rearranging the deck chairs on the **Titanic.**" Even that analogy, though, assumes that all the fat-restricted diet does is prevent heart disease and doesn't do us harm—for instance, make us fatter and more diabetic because of its carbohydrate content.

There is another way to parse these statistics of

population averages, and this is the one the authorities seem to prefer. It is indeed possible that a few of us will die prematurely, perhaps at fifty instead of eighty, as a direct result of elevated cholesterol. If those people eat a cholesterol-lowering diet, they will live significantly longer. But they don't know who they are in advance—nobody does—so we all have to eat the cholesterol-lowering diet for those lucky people to benefit. The rest of us would get no benefit at all. We may even be harmed by such a diet, as many doctors now believe. In 1999 one of the legendary experts in cholesterol research, Scott Grundy of the University of Texas, described this to me as the I-have-to-eat-a-low-fat-diet-for-life-so-my-neighbor-down-the-street-doesn't-get-a-heart-attack scenario. Ninety-nine out of one hundred of us who avoid butter and bacon for a lifetime may well do it for no health benefit whatsoever, even if the conventional wisdom on saturated fat is right.

Physicians who embrace and prescribe LCHF/ketogenic eating believe that these conventional healthy-diet **hypotheses** are refuted daily in their lives and in their practices. After all, many of them and their patients had lived and eaten by these conventional guidelines while they got progressively fatter and sicker (as had I). Some had been vegetarians, even vegans, but LCHF/ketogenic eating was what eventually allowed them to easily lose their excess fat and reverse any progression toward hypertension or diabetes. That's what they directly observed, and that, in turn, is what their patients experience. No faith is necessary to observe or experience these benefits.

Recall what the hundred-plus Canadian physicians wrote in **HuffPost** about their observations, their experiences, when their patients embraced LCHF/ketogenic eating: "What we see in our clinics: blood sugar values go down, blood pressure drops, chronic pain decreases or disappears, lipid profiles improve, inflammatory markers improve, energy increases, weight decreases, sleep is improved, IBS [irritable bowel syndrome] symptoms are lessened, etc. Medication is adjusted downward, or even eliminated, which reduces the side-effects for patients and the costs to society. The results we achieve with our patients are impressive and durable." Physicians who now prescribe these diets commonly say that they rarely if ever prescribe drugs to their patients for blood sugar control or hypertension; rather, they deprescribe, they get patients off medications. That's compelling testimony.

One physician I interviewed put this trade-off in perhaps its starkest perspective. Caroline Richardson is a family medicine doctor at the University of Michigan and a health services researcher who also works for the university's Institute for Healthcare Policy and Innovation. She started her career doing physical activity research and then gradually transitioned into diabetes prevention. For years, she told me, she counseled her patients to follow the Diabetes Prevention Program regimen of low-fat, calorie-restricted diets plus exercise. Most of her patients, though, were extremely obese and half were diabetic. Slowly she shifted into studying and prescribing LCHF/ketogenic eating—typically, after finding out how well a relatively low-carb diet worked for her.

Now Richardson tells her patients to read **Always Hungry?: Conquer Cravings, Retrain Your Fat Cells, and Lose Weight Permanently** by David Ludwig, a physician and professor of nutrition at Harvard, and to study its low-carb recipes. "One thing I love about the low-carb, high-fat diet, which I say again and again to my patients, is it makes you feel better." The situation is similar to that of exercising, she told me. She advises her patients to exercise not because they'll be healthier five years from now. She suggests they do it because they'll feel better now. "When my patients cut out the carbohydrates, every single one comes back saying, 'Wow, I feel like a new person.' And one thing my patients tell me all the time is, 'I don't care if I die in ten years, I feel like crap today, I want to stop feeling like crap today.'"

Dan Murtagh's take on this trade-off is also worth hearing. Murtagh is a general practitioner working in Northern Ireland with a patient population of mostly middle- and working-class families. He told me that when he was in medical school—he graduated in 2002—he heard little discussion about an obesity or diabetes epidemic. By the time we spoke fifteen years later, he was diagnosing a new case of type 2 diabetes in his clinical practice weekly. He became interested in diet and nutrition in 2009 when a patient asked him about the safety and efficacy of a paleo diet.

Murtagh did his homework and went "down the rabbit hole." First he read **The Paleo Diet** by Loren Cordain, the Colorado State University exercise physiologist who did the formative thinking on this way of eating. That led Murtagh to books on LCHF/ketogenic eating. He says the arguments in these books (including mine) made sense to him, so he experimented on himself and then tried it on his patients. "It's all very well waxing on about what you think is going to happen on these diets," he said to me, "but eventually you have to roll up your sleeves and get to work and see what happens."

When I interviewed Murtagh, he told me about several patients whom he had counseled to avoid carbohydrates and to replace those calories with natural (vintage) fats. About one diabetic patient, "not particularly heavy-set," he said, "I don't think **remission** is a strong enough word for what happened to his diabetes." He described another patient, in his early fifties, as "textbook obese": six foot one, 320 pounds, on his way to becoming diabetic, but already with fatty liver disease, gout, and hypertension. Prior to changing how he ate, this patient was taking two medications daily for his blood pressure, another medication for his gout, and another for chronic indigestion and heartburn. After a year of LCHF/ketogenic eating, he had lost upward of 110 pounds and was medication-free.

Surely he was healthier, but Murtagh's medical colleagues who were still bound to the conventional thinking were not sanguine. "I discuss the same patients with them I've discussed with you," he said, and he gets pushback. "I'm thinking, 'Look, you're telling me I should go back to this patient who's lost 110 pounds and got off all his medications, and tell him to go back to eating his bread and cut the fat off his bacon.'"

The fact that LCHF/ketogenic eating produces such remarkable results in the clinic has always represented a tremendous challenge to the conventional thinking on nutrition. It creates an essential conflict, a cognitive dissonance, between two seemingly mutually exclusive definitions of what it means to "eat healthy." Over the last fifty years, healthy eating has conventionally been defined and institutionalized to mean eating fruits, vegetables, whole grains, and pulses in abundance, with plenty of carbohydrates—mostly plants—and minimal animal fats and little or no red or processed meats. The other definition is what many people appear to need to maintain a "healthy" weight: ideally little or no fruit, no whole grains, no legumes or pulses, very few carbohydrates, and plenty of fat, which often translates to plenty of red and even processed meat. How do we resolve the discrepancy? If achieving and maintaining a "healthy" weight requires us to eat an "unhealthy" diet, are we healthier or not?

As clinical experience with these trials has been accumulating, so, finally, has the clinical trial evidence. When I first reported on this subject for that 2002 **New York Times Magazine** article, we were seeing only the very first clinical trials assessing the relative benefits and risks of these eating patterns. These trials informed my decision to take the unorthodox position that I did in the article. Once researchers and authorities in the 1960s chose to believe that all obesity was caused by eating too much and then embraced the notion that saturated fat was a primary cause of heart disease, they did their best to put the entire nation and then the entire world on diets that would hypothetically prevent heart disease. No meaningful research was done on even the short-term effects of LCHF/ketogenic eating. That remained the case through the end of the century. (In the course of my research, I interviewed researchers in Germany who had done clinical trials on LCHF/ketogenic diets through the mid-1980s, then stopped doing them when they decided that the consensus opinion on the dangers of fat must be right, even though that was the opposite of what their own research implied.)

Only at the turn of this century, with the awareness of an obesity epidemic and typically motivated by a personal conversion experience, did physicians begin once again to conduct clinical trials on LCHF/ketogenic eating. In my article, I noted that five clinical trials had recently been completed (albeit not yet published) comparing the LCHF/ketogenic Atkins diet to the kind of low-fat, calorie-restricted (semistarvation) diet recommended then and still by the American Heart Association. The trial participants ranged from overweight adolescents in Long Island, who followed the diets for twelve weeks, to Philadelphia adults whose weight averaged 295 pounds and who followed these diets for six months.

The results of those five studies were consistent. The participants eating the LCHF/ketogenic high-fat diet lost more weight, despite the advice to eat to satiety, than those who ate the AHA-recommended low-fat, low-saturated-fat diet. Moreover, their heart disease risk factors showed greater improvement. In other words, the results of these trials were the opposite of what physicians and medical researchers would have predicted. And this is what I reported.

Since then, as of the spring of 2019, close to one hundred, if not more, clinical trials have published results, and they confirm these observations with remarkable consistency. The trials are still incapable of telling us whether embracing LCHF/ketogenic eating will extend our lives (compared to other patterns of eating the authorities might recommend), but they continue to challenge, relentlessly, the conventional thinking on the dangers of high-fat diets, and they tell us that in the short term, this way of eating is safe and beneficial.

Following LCHF/ketogenic eating for the duration of these clinical trials (at most two years), or at least being assigned to eat that way, results in equal or greater weight loss than any eating pattern to which it has been compared, and that happens without requiring the study participants to count and consciously restrict calories. And the benefits to health are clear. As with the first five studies and the clinical experience, virtually all measures of metabolic health, all risk factors for heart disease and diabetes, improve with LCHF/ketogenic eating. Along with achieving a healthier weight, the study participants became healthier overall, and they become healthier than the participants who are counseled to eat conventionally "healthy," even calorie-restricted, diets.

One particularly compelling trial was recently completed at Indiana University, led by Dr. Sarah Hallberg, working with a San Francisco–based start-up called Virta Health that was founded by Steve Phinney and Jeff Volek. Hallberg and her colleagues counseled patients with type 2 diabetes to follow LCHF/ketogenic eating. They provided 24/7 guidance from health coaches and physicians to address any issues that arose and to help them stick with it. Even in participants with type 2 diabetes, LCHF/ketogenic eating consistently produced the kind of results that we should now expect: These people were not counseled to eat less, yet they experienced significant weight loss. Their cardiovascular risk factors improved significantly. And perhaps most important, many of the 262 participants assigned to the LCHF/ketogenic eating arm of the IU/Virta Health trial had their diabetes effectively go into remission. Blood sugar control improved even while they discontinued their blood sugar medications, including insulin. ("Insulin therapy was reduced or eliminated in 94% of users," the Virta Health team reported.) Blood pressure also improved, and so blood pressure medications were stopped as well.

In June 2019 Hallberg and Virta Health published a paper on how two years of LCHF/ketogenic eating had influenced heart disease risk factors in its subjects. The bottom line was that twenty-two of twenty-six established risk factors improved (compared to what these physician researchers call "usual care"), three remained unchanged, and only one—LDL cholesterol—on average got worse. When the Virta Health research-ers worked out the numbers for what's called the "aggregate atherosclerotic cardiovascular disease risk score," a measure of ten-year risk of having a heart attack developed by the American College of Cardiology and the American Heart Association, the Virta Health patients decreased their risk of having a heart attack by over 20 percent, compared to the usual treatment program for diabetes and all the drug therapies typically prescribed. Even with the rise in their LDL cholesterol, these pa-tients got significantly healthier, as did their hearts.

So here's yet another way to ask the critical ques-tion: Can a pattern of eating that has so many beneficial effects be unhealthy because it contains considerable saturated fat or allows for the con-spicuous consumption of a processed meat like bacon?* In one of my favorite Rachelle Ploetz Instagram postings, she noted that her friends never criticized her diet when she weighed 380 pounds, but having switched to LCHF/ketogenic eating and lost by then 120, they would often express concern about how much bacon she was eating, as

* Not that bacon or meat of any kind is required in LCHF/ketogenic diets, but as foods that contain (essentially) only protein and fat, they can be eaten freely. With the exception of avocados, olives, and vegetable oils, plant-based foods come with carbohydrates as a sig-nificant source of available energy. Lower-carb, higher-fat versions of plant-based diets can be consumed, but they take significantly more thought and work and may or may not be as effective. These are discussed in Chapter 16.

though the dangers of eating bacon regularly out-
weighed the benefits of losing 120 excess pounds
with relative ease.

The definitive evidence to answer this question
does not exist. It may never exist. But it's hard to
imagine that a way of eating that makes people
so much healthier in the short run, that can even
reverse diabetes, which is considered a progressive
chronic disease—one that only gets worse as time
goes by—will harm us in the long run. The authorities are willing to think in terms of hypotheti-
cals and hold on dearly to their cherished beliefs.
Those beliefs have already failed us. We have to
take the gamble and leave them behind.

Our institutional condemnation of dietary fat
and the wisdom behind prescribing diets by hypothesis would be more understandable if the
evidence to support these hypotheses were indeed
compelling. I don't believe it is. Just as the evidence has inexorably accumulated over the years
supporting the observation that LCHF/ketogenic
diets make us healthier, the evidence supporting
the idea that saturated fat is deadly and that we
should all eat low-fat diets has been fading, despite
the best efforts of the orthodoxy to prop it up. The
more research that's been done, the less compelling
it becomes. This is always a bad sign in science
and a persuasive reason to believe that a theory or
a belief is simply wrong. Outside mathematics, it's
impossible to prove anything definitively one way

or the other. Evidence always exists to support reasonable hypotheses (and even some unreasonable ones), because studies will always be done that get the wrong answer or that are interpreted incorrectly. That's why I suggest we follow the trends.

The better scientists and philosophers of science have been advising this approach at least since Francis Bacon (his name, of course, is only coincidental) pioneered the scientific method four hundred years ago: The way to judge the viability of a hypothesis is to judge whether the evidence has grown significantly stronger with time. As Bacon suggested, you can tell what is **not** correct in science—what he called "wishful science," which is based on fancies, opinions, and the exclusion of contrary evidence—because these are the propositions that "have stuck fast in their tracks and remained in virtually the same position, without any noticeable development; rather the reverse, flourishing most under their first authors, but going downhill ever since."

The dietary fat–heart disease hypothesis, the one on which we base our anxieties about eating saturated fat, should be a case study in this kind of downhill progression. In 1952, while acknowledging he had no meaningful evidence to support his proposition, Ancel Keys suggested Americans should eat one-third less fat than they were at the time if they wanted to avoid heart disease. In 1970, still without hard clinical trial evidence, the American Heart Association recommended low-fat diets for everyone in America literally old enough to walk. In 1988, after the publication of two hundred-million-dollar-plus clinical trials, the results of which happened to be contradictory, followed by what one NIH administrator later described to me as a "leap of faith," the U.S. surgeon general was blaming two-thirds of the two million yearly deaths in the United States on the overconsumption of fat-rich foods, and maintaining that the "depth of the science base" was "even more impressive than that for tobacco and health." That report was part of a concerted public relations campaign by the federal government to do all it could (apparently with the best of intentions) to get us to fear eating any fat that didn't come from vegetable sources. It worked. That's why we bought into the idea that we should avoid eating saturated fat if at all possible. Animal fat consumption in America went down; plant oil consumption went up.

Now, thirty years later, the most recent **unbiased** review of this evidence—from the Cochrane Collaboration, an international organization founded to do such impartial reviews—concluded that clinical trials have failed to demonstrate any meaningful benefit from eating low-fat diets and so, implicitly, any harm from eating fat-rich foods. The Cochrane review described the evidence as only "suggestive" that avoiding saturated fat specifically might avert a single heart attack, and said it's even "less clear" whether this would lengthen anyone's life.

Despite its prominent role in pushing the anti-fat frenzy, the American Heart Association recently acknowledged (in an otherwise biased assessment) that its conception of healthy low-fat eating gets support primarily, still, from the ambiguous results of a handful of poorly done trials that **all** date to the 1960s and '70s, and that if this murky evidence is to take precedence, the results of later studies, including the enormous (49,000 participants) and exorbitantly expensive (half a billion dollars at least) Women's Health Initiative, have to be ignored or rejected as inadequate. Of course, the hundred-some trials consistently finding that LCHF/ketogenic eating makes us healthier, despite being saturated-fat-rich, also refute the idea that we should be listening to the authorities. For the past half century, evidence supporting the idea that the saturated fat in our diet is a cause of heart disease and premature death has simply been eroding away.

The notion (i.e., hypothesis) that fat-rich foods cause cancer has had similar setbacks. In 1982 this proposition was considered so likely to be true that the National Academy of Sciences published a report—**Diet, Nutrition, and Cancer**—recommending that to prevent cancer, Americans cut fat consumption from 40 percent of our calories, as we were then eating, to 30 percent. It asserted that the evidence was so compelling, it "could be used to justify an even greater reduction." This is also what health-conscious people grew up believing and were taught. By the mid-1990s, though, the experts who assembled a seven-hundred-page report on this question for the World Cancer Research Fund and American Institute for Cancer Research—**Food, Nutrition and the Prevention of Cancer**—could find neither "convincing" nor even "probable" reason to believe that fat-rich diets were carcinogenic. When I interviewed Arthur Schatzkin, chief of the nutritional epidemiology branch of the National Cancer Institute, in 2003, he described the evidence from clinical trials designed to test this dietary fat–cancer hypothesis as "largely null." In short, the proposition that fat caused cancer had also gone steeply downhill with further study, but our fear of fat did not go with it.

As for the idea that a healthy diet must be mostly plants, that it must include fruits, vegetables, whole grains, pulses, and legumes, we don't have even the ambiguous 1960s-era studies to support it. We have no meaningful clinical trial evidence to support this idea, as Michael Pollan infers in **In Defense of Food,** the book that brought us the mantra "Eat food. Not too much. Mostly plants." What we have instead, he notes, is the idea that people who eat a lot of plant foods tend to be healthier than people who eat the standard

American diet (given the appropriate acronym SAD), that is, who eat at fast-food restaurants and buy the packaged, highly processed, sugary foods in the supermarket that Pollan aptly calls "foodlike substances," food that health-conscious people naturally avoid. More than anything, says Pollan, we have the simple fact that virtually all nutritionists believe eating mostly plants is a good idea. In the very contentious world of nutritional beliefs, he says, this is something on which they can all agree.

Yet they believe this, and Pollan argues for it, not because they have compelling experimental evidence (i.e., clinical trial results) that it is true, and not because they've seen obese and diabetic patients switch from omnivorous or meat-rich diets (without sugar and foodlike substances) to mostly or all-plant diets (without sugar and foodlike substances) and get healthier for doing so, but because they, well, all seem to believe it. This is what cognitive psychologists would call a "cascade" or "groupthink," and it's exceedingly common in this kind of soft science. It's even common in the harder sciences—physics, for instance, where the Nobel laureate Louis Alvarez called it "intellectual phase lock." People believe something because people they respect believe it, and if they're doing research, they report what they're supposed to find, and they see what they expect to see, whether it's really there or not.

Eating mostly plants, in other words, **just seems** right to those who recommend it to us. It seems right, in part, because we've been hearing it our whole lives. It's what my health-conscious mother was teaching me in the 1960s every time she told me to eat my vegetables (if it wasn't green or cauliflower, in her worldview, it was not a vegetable) and suggested that too much red meat would cause colon cancer. I'm now badgering my children to eat their green vegetables, even if I primarily believe they should because that's what my mother taught me. Eating mostly plants may be better for the environment than the alternatives, and better for the animals that won't be killed prematurely and eaten for our pleasure.* When epidemiological surveys look at what healthy, health-conscious people eat, not surprisingly this mostly plants wisdom wins out. Health-conscious people have not been sitting down to breakfasts of eggs and bacon every morning, because they've been told eggs and bacon will kill them. They're drinking kale-almond smoothies with their low-sugar granola because that's what they've been counseled, no matter the weakness in the underlying evidence. Shouldn't we all?

The answer, once again, is probably not. The last

* Sophocles counsels at the end of **Oedipus Rex** that we should look upon that last day always and count no mortal lucky or happy until he (or she) lives his last day without pain. If the same is true of animals, then this assumption, too, is questionable.

thirty years of medical research have resulted in a sea change in our understanding of heart disease risk factors and their relationship with obesity, diabetes, and the condition we discussed earlier called insulin resistance. A critical factor in the pushback against LCHF/ketogenic eating has always been the belief that the animal fat content will cause premature heart disease—the "artery-clogging" saturated fat argument. Most people believe butter and bacon and full-fat dairy products are deadly because we've been taught that these foods high in saturated fats will raise our cholesterol, specifically the cholesterol in LDL particles known as the "bad" cholesterol, and that this will lead to premature death from a heart attack.

One of many problems with this way of thinking is that it focuses all dietary attention on one disease state, heart disease, and one biological entity, LDL cholesterol. This is at best misguided 1970s-era medical science. While physicians have been taught to believe it with dogmatic certainty, and a large proportion still do, the scientific understanding has evolved over the years, as scientific understanding has a way of doing.

While LDL does seem to play a role in the atherosclerotic process, it's not the cholesterol in the particle that's the active player but rather the LDL particle itself and specifically the number and maybe the size of particles in circulation. Public health and medical authorities have slowly come to accept what research and physician iconoclasts had argued as early as the 1960s, that heart disease is a complex process and the end result of a metabolic disruption that manifests itself throughout the human body. We cannot ascertain whether we will live a long and healthy life from a single number and a single biological entity. (The measures that are best at doing that, in any case, are far better indicators than LDL cholesterol.) For most of us, the primary sign that we're at high risk of heart disease or premature death from any chronic disease, including cancer, is not whether our LDL cholesterol is elevated, but whether we have the cluster of metabolic disorders now known as metabolic syndrome, which itself seems to be a consequence or manifestation of insulin resistance.

Physicians are instructed to diagnose metabolic syndrome if their patients have at least three of five characteristic signs. The most important, the one physicians are told to look for first, is whether the patient is getting fatter, specifically above the waist. In this sense, the metabolic syndrome concept is, perversely, a direct descendant of Ancel Keys's thinking and observations in 1960 that the people most likely to get heart attacks and die prematurely are fat middle-aged men, those fat men that Keys was so ardently imploring to "think." Some heart specialists were referring to these men as "fat cardiacs" even a century ago. Keys and the medical community became obsessed with dietary fat and cholesterol as the key to solving the fat-man-heart-attack connection and so focused all attention on LDL cholesterol and dietary fat. But other researchers—at Stanford, Yale, and Rockefeller universities in the United States and at Queens Elizabeth College and Queen's University in Belfast, among others—focused on carbohydrates and their effect not just on insulin and elevated blood sugar but also on blood pressure and "blood lipids," in particular HDL cholesterol (the "good cholesterol") and triglycerides (one form in which fat is found in the circulation). This is what Edwin Astwood was referring to in his 1962 lecture when he observed that the disorders associated with obesity—"particularly those involving the arteries"—closely resemble those of type 2 diabetes, implying "a common defect in the two conditions."

By the late 1980s, as the National Institutes of Health, the Surgeon General's Office, and even the National Academy of Sciences in the United States—not to mention the National Health Service in the United Kingdom—were convincing us all to avoid fat and eat carbohydrates, researchers led by the late Stanford University endocrinologist Gerald Reaven began to convince first diabetes specialists and then eventually cardiologists that their patients should be worried less about LDL cholesterol than about metabolic syndrome. It was metabolic syndrome, these physician researchers argued, that was the manifestation of the fundamental physiological disruption that would eventually kill their patients (and us). This is what journalists are referring to when they write, as Trymaine Lee, an NBC correspondent, recently did, that "obesity and high blood pressure [are] key contributors to heart disease." Lee was writing about his own near-fatal heart attack at age thirty-eight. Obesity and high blood pressure are manifestations of metabolic syndrome; they go hand in hand.

The revelations about metabolic syndrome can be understood if we think of obesity, diabetes, heart disease, hypertension, and even stroke all as consequences of the same disruptive force: disordered insulin signaling, poor blood sugar control, and all the metabolic and physiological disruptions, including systemic inflammation, that then occur. All these conditions are intimately associated. Those who have obesity are at high risk of type 2 diabetes, and most people with diabetes are overweight or obese. They're all likely to get heart disease (as Astwood noted), but those with diabetes are at the highest risk, and they all tend to have high blood pressure. Medical textbooks refer to obesity, diabetes, heart disease, gout, and stroke (cerebrovascular disease) as "hypertensive" disorders, meaning high blood pressure is common in all of them. Additionally, all these disorders associate with these abnormalities in blood lipids, specifically low HDL cholesterol and high triglycerides (and high LDL particle number, but not high LDL cholesterol).

These risk factors are the diagnostic criteria of metabolic syndrome. Individually, each of these factors is associated with an increased likelihood that you'll have heart disease: As your waist circumference expands, your risk of heart disease goes up. As your blood pressure elevates, so does your risk for heart disease, and stroke as well. The worse your blood sugar control (glucose intolerance), the more likely you are to be diabetic, and the more plaque deposition you're likely to have in your arteries. In 1930 Elliott Joslin, the leading U.S. authority on diabetes, observed that "every other diabetic now dies of arteriosclerosis," and the situation hasn't changed much since then. The arteries of a sixty-year-old with untreated diabetes will look like the arteries of a ninety-year-old who doesn't have the disease. Finally, the medical community has known since 1977 (if not twenty years earlier) that low HDL cholesterol is a far better predictor of heart disease than high LDL cholesterol, many times more likely to be regrettably right, and that high triglycerides are at least as predictive as high LDL. The likelihood is that when you have a heart attack, metabolic syndrome will be the reason, not your elevated LDL cholesterol.

If you have metabolic syndrome, it means you're sliding down the slope from health to chronic disease, and the first obvious sign is that you're getting fatter or you've got high blood pressure. According to Centers for Disease Control (CDC) statistics, one in three Americans has metabolic syndrome. But that proportion includes children, in whom it is relatively rare. The older we get and the fatter we get, the more likely we are to have metabolic syndrome, to be insulin resistant. Among adults over fifty, one in two have it. If you're reading this book to help bring your weight under control (and particularly if you're male), it's a good sign that you either have metabolic syndrome or are going to get it.

All these physiological disturbances that characterize metabolic syndrome, all the risk factors that physicians are told to look for to diagnose metabolic syndrome, are linked directly to the carbohydrates we eat, not to the fat. If you have metabolic syndrome, it's the quantity and quality of carbohydrates you're eating that are slowly shortening your life. Saturated fat is not responsible. Both clinical trial data and clinical experience tell us that this body-wide disruption of metabolic syndrome—the disruption that appears to begin with insulin resistance and so elevated levels of insulin and poor blood sugar control—is normalized or corrected by removing the carbohydrates from the diet and replacing them with fat. That's the twenty-two of twenty-six risk factors that improved in the Virta Health trial.

All this—what happens to the human body when blood sugar and insulin move in and out of healthy ranges—can be explained by textbook medicine. By this I mean that the beneficial effects observed when patients or clinical trial participants restrict carbohydrates and replace them with fat are what medical textbooks tell us should happen. Eating fewer carbohydrates, for instance, will, by definition, result in lower blood sugar, at least in the short term after a meal. This almost has to be beneficial, considering it's high blood sugar that causes many of the deleterious side effects of diabetes. Researchers have known, at least since the 1970s, that carbohydrate consumption lowers the apparently beneficial HDL cholesterol compared to eating fats, and that it raises triglycerides as well. Their understanding of how the liver processes these "lipids" and lipoproteins explains why.

As for blood pressure, insulin induces your kidneys to hold on to sodium. (Salt is sodium chloride, and the sodium is the player here.) This is one of the many things insulin does. When your insulin levels are high, your kidneys retain sodium rather than excreting it in urine. Now blood pressure will increase as your body retains water to keep the sodium concentration in your circulation constant. When the medical authorities blame hypertension and high blood pressure on eating too much salt, they're thinking of the same mechanism—increasing the sodium concentration in the circulation leads to more water being retained and higher blood pressure—but typically simplistically. They're putting the blame on consuming too much salt—a behavioral problem or maybe the food industry's fault for oversalting processed food—rather than on excreting too little, which results from chronically elevated insulin levels and insulin resistance. Lowering insulin by avoiding carbohydrates and replacing them with fat reverses this sodium-retention phenomenon, and so blood pressure should drop with LCHF/ketogenic eating, as it typically does.

Once again, knowing the history of nutrition science makes the fact that orthodox medicine has ignored this connection all that much more disturbing. As early as the 1860s, the German biochemists who pioneered the science of nutrition were commenting that carbohydrate-rich diets elevated blood pressure and fat-rich diets did not. In the 1970s Harvard researchers came to understand the role of insulin in this process. By then, though, we were all being told that high blood pressure was caused by eating too much salt, another speculative hypothesis that continues to suffer from a dearth of experimental, clinical trial evidence. It was embraced nonetheless. It sounded right, and so the authorities believed it. We believed it because they did, and we never let it go.

Meanwhile business in blood pressure medications

boomed—tens of billions of dollars a year worldwide—and the carbohydrate-insulin–blood pressure connection was relegated to the textbooks. Like most things insulin-related, it is assumed to have no relevance to anyone other than maybe those with diabetes. By the mid-1990s diabetes textbooks, such as **Joslin's Diabetes Mellitus,** described chronically elevated levels of insulin as likely to be "the major pathogenic defect initiating the hypertensive process" in patients with type 2 diabetes. Patients with type 2 diabetes are just further down the metabolic syndrome spectrum than the rest of us, but this idea that chronically elevated levels of insulin might be the pathogenic defect initiating the hypertensive process in the rest of us was not thought to be relevant. But it is, though, certainly to those of us who want to be lean **and** healthy.

Nutritional authorities (or at least those quoted in the media) still argue with the same dogmatic assurance as ever that they've always been right and so their credibility should not be doubted, but the conventional thinking on nutrition and the nature of a healthy diet has clearly changed considerably in the past twenty years. The slow, relentless accumulation of clinical trial and clinical evidence supporting what I'm arguing for in this book, and what thousands of physicians have now come to believe, has had an effect, which is how science is supposed to work.

Twenty years ago, when I first began reporting on this subject, the conventional wisdom was that **the only way** to lose weight was to consciously restrict calories (or exercise more); that diets that prevented heart disease **had to be** low in fat; and that LCHF/ketogenic eating was deadly. Now, with the notable exception of Katz and Bittman in their hyperbolic moods and **U.S. News & World Report** (where Katz has played a significant role in the authoritative committee), proponents of the orthodoxy in the media are typically arguing or defending a much different position: that calorie-restricted and/or low-fat diets are **as good** or **as healthy** as LCHF/ketogenic eating, about which there is nothing special. These conventional eating experts want people to know we still have a choice when it comes to weight loss (therefore these experts weren't completely wrong, only partly). The informed argument is no longer that LCHF/ketogenic eating will shorten our lives but that other ways of eating may work just as well. The implication is that these other eating patterns aren't as radical, making them easier to sustain and surely less of a risk.

A handful of prominent physicians and nutritional authorities will still actively argue—as they do, for instance, in the Netflix film **What the Health**—that the healthiest way to eat for all of us is to minimize animal fats and animal products. Not just to consume **mostly** plants but perhaps to eat **only** food that is plant-based, vegetarian, or even vegan. But these physicians or researchers have not compared these two approaches—whether in their own clinics or in clinical trials—to conclude that mostly plant diets work better for their patients or that LCHF/ketogenic eating does harm. (A reminder: The relevant trials that can do this reliably don't exist.) These physicians, nutritionists, and even epidemiologists surveying populations have strong beliefs that mostly or all-plant diets are beneficial, which may be valid. But that tells us (and them) nothing about the relative benefits or harms of LCHF/ketogenic eating. These physicians don't know, in effect, whether their patients would do better or worse abstaining from carbohydrate-rich foods specifically, rather than from animal products. They're only guessing. Those who insist so vehemently that these diets are dangerous do so not because they have clinical experience to that effect, and not because they're familiar with the clinical research literature, but because they don't.

So is it safe? Can you follow an LCHF/ketogenic eating plan indefinitely without fearing that you're slowly killing yourself? The existing evidence says that if you have metabolic syndrome, if you're getting fatter or are already obese, if you're prediabetic or already diabetic, avoiding carbohydrate-rich foods and replacing them with fat may be the single healthiest thing you can do for yourself. That's why so many physicians have now become evangelists.

No one can guarantee what happens in the long run. The evidence to do so, as I've said repeatedly, doesn't exist and may never exist. Anyone who makes an ironclad guarantee for any way of eating—that one diet will assuredly make you live longer than others—as Gladwell suggested and I tend to agree, is probably selling something (although perhaps with the best of intentions).

Over the years, both the media and the research community have fallen into the habit of discussing the constituents of healthy diets in terms of the benefits they may confer. Eating abundant fruits and vegetables, as a recent **New York Times** article said, "can promote health," as though these foods contain indispensable ingredients that work to make us healthy and keep us healthy. By this logic, the more fruits and vegetables in a diet, the better. This may be true, but the only way we can get some reliable information is by adding them to our diets or taking them away and seeing what happens. Do we get leaner? Do we get healthier? Do we feel better or worse?

A more useful way to discuss the pros and cons of dietary changes, as implied earlier in this chapter and by Geoffrey Rose's observation about natural and unnatural factors, is in terms of how well they do at removing whatever it is that makes us ill, while keeping the essential fats, minerals, and vitamins that we know reliably are necessary for health. (If we eat in a way deficient in these fats, minerals, and vitamins, we get deficiency diseases.) By this standard, we know that when carbohydrates are removed (including fruits and starchy vegetables) and replaced with fat, people get leaner and healthier. What was wrong with these folks has seemingly been corrected by the simple removal of nonessential constituents of the diet.

As such, LCHF/ketogenic eating can be thought of as working to correct our health rather than improve it. I'm proposing that's how we should think about it.* A diet that restricts carbohydrates and replaces those calories with fat **corrects** your weight by lowering it. It **corrects** your blood pressure by lowering it. It **corrects** your inability to control your blood sugar. It's not the equivalent of taking a pill that will make you healthy; rather, it removes what makes you unhealthy, replaces those calories with a benign macronutrient (fat), and in so doing, fixes what ails you. These corrections are noticeable in real time, by patient, by physician, and by any individuals who try this approach on their own.

---

* This is another concept for which I cannot take credit. That goes to my friend Bob Kaplan, who is not an academic researcher but an amateur (like me, in that sense). He owns a string of health clubs in the Boston area, has a formal education in exercise physiology, and has made it his life's pursuit to understand the relevant science. He's done as good a job of it as anyone I know.

The gamble is that improving health in the short run will lead to improvements in the long run. We're gambling that if something happens in the future, if a symptom of ill health develops, we can experiment with how we're eating to see if that's the cause, then fix it appropriately. We're taking our health into our own hands. There are no ironclad guarantees, though. There never are.

In considering the question of what's safe and what isn't, one more vitally important aspect has to be taken into account. It's no longer just our health that we're worried about or our children's—it's the planet's. So we must ask if LCHF/ketogenic eating is justifiable if it means increasing your "climate footprint" compared to alternatives. Given what may be a trade-off between humanity's future and your own health (and that of your children), how do you decide?

In the last few years, the conventional wisdom has emerged that eating animal products results in a greater contribution to greenhouse warming of the planet than does eating plants. Because we worry with good reason that global warming is a major threat to planetary health and humanity's future, we believe we should do whatever we can personally to mitigate it. This has led newspapers to publish analyses of "how to shop, cook and eat in a warming world," as **The New York Times** did in April 2019, and to suggest that the fewer animal products we consume (and certainly the less beef, lamb, and dairy, as these seemingly have the greatest climate footprints), the healthier the planet will be.

This may indeed be true. While acknowledging that livestock can be raised in ways that are relatively climate friendly and much of it is (in the United States, for instance, more, say, than in Brazil), the implication is that the most climate-friendly eating pattern is one that omits these foods—a vegan diet—and that that's how we should eat. For those who don't think they can become a vegan, the **Times** suggests, then "another approach would be to simply eat less meat and dairy, and more protein-rich plants like beans, legumes, nuts and grains."

The problem, of course, is that this thinking once again assumes that the conventional healthy diet—or even an unconventional and arguably unnatural diet, per Geoffrey Rose's thinking, like the vegan diet—is indeed healthy for all of us. It builds on a foundation of the bad science in nutrition research of the past fifty years, and it shows little concern for the absence of clinical trials that might actually test it. It's also the lean person's perspective. If those of us who are predisposed to be insulin resistant, obese, and/or diabetic in the modern food environment get fat or stay fat eating beans, legumes, and grains, we have a conflict that must be resolved.

It's certainly possible to eat a vegan or vegetarian

LCHF/ketogenic diet, and many people now do. Whether it is a healthier option for some of us (rather than for the environment) in the long run than LCHF/ketogenic eating with some or even mostly animal products is an open question. I'm skeptical (as is my nature). Without the clinical trials, the only evidence we have on which to base our conclusions is how our weight and health status responds to these eating patterns. As we try to do what we can for the environment, the planet, and our future, we'll have to take into consideration what we have to eat to remain healthy and how important that is to us. Until we know the trade-offs, both personally and as a society, it may be a costly mistake, regrettably, to assume that a way of eating that is healthiest for the planet is healthiest for us.

13

Simplicity and Its Implications

Everything should be as simple as possible, but no simpler.*

The message should be straightforward: Carbohydrate-rich foods are fattening. Or to complicate it slightly such that naturally lean people might more likely understand: For those of us who fatten and particularly those who fatten easily, it's the carbohydrates that we eat—the quantity and the quality—that are responsible. The relevant mechanism appears to be simple, as well: Carbohydrate-rich foods—grains, starchy vegetables, and sugars—work to keep insulin elevated in our circulation, and that traps the fat we eat in our fat cells and inhibits the use of that fat for fuel.

That's what the obesity research community

---

* Thanks to Albert Einstein for this thought, although he was talking about scientific theories, not how to eat, and this is probably a simplification of what he actually said, not a direct quote.

should have been trying rigorously to resolve or refute for the past sixty years. That's what I'm assuming is true because of the reasons and the evidence discussed. That's what we have to keep in mind as we think about how to eat.

This simple truth about carbohydrates seems so hard to understand because we've been trapped in a context of naïve conventional wisdom—eat less or not too much, avoid fat and saturated fat, eat mostly plants—which in turn spawned the fad diet phenomenon I've discussed throughout. By relegating the reality of dealing with obesity and overweight to practicing physicians, the nutritional authorities almost guaranteed a future in which reality and the straightforward steps required to overcome obesity-related disorders would be hard to discern.

When these physicians wrote self-help books about what they had learned during their conversion experiences, books that contradicted the conventional low-fat, eat-less, mostly plant wisdom, they had to say something new, something different from the physicians turned authors who preceded them. That's the nature of publishing. It's hard to sell a diet book or website that advises people to eat precisely as others have advised in the past, although for the most part a large proportion of these books are merely minor variations on this theme.

With each new addition to the diet book

literature, the focus of discussion narrowed to what the books added to the baseline advice, typically what we **should** eat to attain a healthy weight rather than the simple message about which foods are fattening (to us) and that we should avoid. Discussions of paleo versus keto versus South Beach versus the Zone or even versus Weight Watchers and Jenny Craig or the dietary vehicle of Oprah's latest weight-loss achievement focused on the subtle ways these approaches differ rather than what they all have in common: the advice to avoid or mostly avoid, at the very least, refined grains and sugars. As diet guides struggle to give added value and find a new way to sell an old message— one that still desperately needs selling, offering to hone our health to some hypothetical fine edge or allow us to have the longest possible health span or even mind span (avoiding dementia and staying sharp as we age)—they verge further into the speculative, maybe-right-probably-wrong research literature and away from reliable knowledge.

The simple and reliable advice is the same as it has been for the better part of two hundred years. It dates back at least to Jean Anthelme Brillat-Savarin in 1825 and **The Physiology of Taste,** which has never been out of print, an accomplishment that very few nonfiction books can claim after nearly two centuries. Brillat-Savarin got it as right as anyone. He had his own conversion experience, just as fad diet book authors typically do, and he wrote about it. He spent thirty years struggling with his weight—he called his paunch his "redoubtable enemy"—and eventually came to what he considered an acceptable standoff. He did so only after digesting the message "of more than five hundred conversations" he had held with "dinner companions who were threatened or afflicted with obesity." In every case, he wrote, the foods they craved were breads and starches and desserts.

As a consequence, Brillat-Savarin considered it indisputable that grains and starches were the principal cause of obesity*—along with a genetic or biological predisposition to fatten easily, which not everybody has—and that sugar exacerbated the fattening process. He lived in a time, though, when sugar was still a luxury for the wealthy, and sugary beverages were exceedingly hard to come by, at least compared to their ubiquity a century later. So he focused his advice on starches and flour, assuming that abstinence from flour would imply abstinence from sugar, since sugars back then came predominantly in baked goods, pastries, and desserts.

Brillat-Savarin acknowledged that those who wished to reduce their weight needed something more than just the usual advice to "eat moderately" and "exercise as much as possible." The only infallible system, he said, had to be diet, and

---

* Brillat-Savarin was confusing association with causation here.

that diet had to remove the cause of the excess body fat:

Of all medical prescriptions, diet is the most important, for it acts without cease day and night, waking and sleeping; it works anew at every meal, so that finally it influences each part of the individual. Now, an anti-fat diet is based on the commonest and most active cause of obesity, since, as it has already been clearly shown, it is only because of grains and starches that fatty congestion can occur, as much in man as in the animals; in regard to these latter, this effect is demonstrated every day under our very eyes, and plays a large part in the commerce of fattened beasts for our markets, and it can be deduced, as an exact consequence, that a more or less rigid abstinence from everything that is starchy or floury will lead to the lessening of weight.

Brillat-Savarin even went so far as to imagine his readers complaining that more or less rigid abstinence from everything that is starchy or floury meant no longer eating the foods they craved. In other words, his readers then might be much like readers now. "In a single word he [Brillat-Savarin] forbids us everything we most love," he wrote, "those little white rolls from Limet, and Achard's cakes, and those cookies . . . and a hundred other things made with flour and butter, with flour and sugar, with flour and sugar and eggs! He doesn't even leave us potatoes, or macaroni! Who would have thought this of a lover of good food who seemed so pleasant?" Brillat-Savarin's response was simple (although I'm bowdlerizing the translation for the more sensitive times in which we live): Then eat these foods and get fat and stay fat!

For many or most of us, this logic offers little or no escape, and as Brillat-Savarin said, the conclusion can still be deduced as an exact consequence. If carbohydrate-rich foods make us fat, then we have to deprive ourselves of the pleasure of their eating if we want to avoid this fate or possibly reverse it. But then, as Brillat-Savarin also noted, these restrictions left plenty to eat and as much of it as desired, which meant meals could be consumed that were still plenty tempting but not fattening.

In the early 1860s, a formerly obese London undertaker named William Banting published multiple editions of the first internationally best-selling diet book. They sold so widely and so well that in some nations the word for "diet" is still a variation on "banting." Banting, too, had a conversion experience, and he discussed it. He, too, had struggled for decades with his weight before being convinced, in his case, by a London physician, to avoid sugars, starches, and grains, and thereafter he effortlessly slimmed down. The pamphlet he subsequently wrote triggered such an uproar that

The Lancet, a British medical journal, wrote two editorials on the approach. The first one derided Banting for not being a physician himself and suggested he mind his own business. (I can relate.) The second, five months later, took a more balanced perspective and made the point that a "fair trial" was needed to ascertain if "the sugary and starchy elements of food be really the chief cause of undue corpulence."

That's the simple issue, as defined reliably by an editor of a medical journal 150-plus years ago. It's not whether one diet somehow works better than another, or whether a calorie is a calorie (as this subject is often discussed and debated), or whether one diet generates a "metabolic advantage" compared to another. The issue is whether the sugary and starchy elements of the diet are the chief cause of undue corpulence—why we get fat. If they are, as textbook medicine has implied for fifty years, then those are the foods we can't eat.

The implications are also relatively simple. The more carbohydrate-rich the food and the easier those carbohydrates are to digest, the greater the blood sugar and insulin response, and the more fattening they are likely to be. And the greater the sugar content, as Brillat-Savarin suggested, the more fattening.

While starches and flours are absorbed into our circulation primarily as the carbohydrate glucose, the stuff of blood sugar, the sugar in our diet (technically sucrose or high fructose syrups), the sweet stuff, has a different chemical composition and, for this reason, does its damage via a different mechanism. Sucrose is a molecule of glucose bonded to a molecule of another carbohydrate called fructose. Fructose is the sweetest of the carbohydrates, and it's why sugar is as sweet as it is, and why fruits, containing a little sugar and a little fructose, are also sweet when ripe.* When we consume these sugars, the glucose enters the circulation, becomes blood sugar, and stimulates an insulin response, but the fructose mostly doesn't. It's metabolized first in the small intestine and then the liver. These organs, the liver particularly, are then tasked with the job of metabolizing an amount of fructose, day in and day out, which they are apparently ill-equipped to do.

Our livers would be easily capable of metabolizing the trickle of fructose that they would have encountered during the few million years that preceded the coming of agriculture about ten thousand years ago: a little sugar, a little fructose, seasonally, in fruits, bound up in fiber, slow to digest (and not necessarily ripe fruits at that). Our livers might have had to deal with the fructose

---

* High-fructose corn syrup, as we consume it most commonly, is a mixture of 55 percent fructose molecules and 40-plus percent glucose and a few other carbs thrown in as well. For our purposes, it's just another version of sugar, so when I say sugar or sugars, I'm speaking of sucrose and high-fructose syrups.

from honey as well. After the twelfth century, depending on where our ancestors lived and their wealth, the trickle increased very slightly as refined sugar, now separated from the fiber that slowed its digestion and absorption, was first imported from the Middle East into Europe. Then the Industrial Revolution came about, and the beet sugar industry was launched to join the cane sugar industry, and the trickle turned into a flood. In the late 1970s, the corn refiners got into the game with high-fructose corn syrup, and the flood of sugar rose even higher; some variant of sugar was consumed in huge amounts daily by all, from breakfast to postdinner desserts, drinks, and snacks.

From the early years of the nineteenth century to the very tail end of the twentieth, average per capita sugar availability (how much the food industry makes available for our consumption) increased in the United States more than thirty-fold: from the sugar equivalent of a single twelve-ounce can of Coca-Cola every **week** to that of more than five cans every **day,** for everyone, from newborns to centenarians.

Like any device tasked to do a job it isn't designed to do, the liver does a poor job of metabolizing this daily flood of fructose. Liver cells use as much of the fructose as they can to generate energy, but they convert the rest, the excess, to fat. Reasonably reliable research suggests that this fat is trapped in liver cells, leading to a condition known as nonalcoholic fatty liver disease, which is associated with obesity and diabetes and is also becoming an epidemic in the modern world. Some very good biochemists think that the backup of fat in these liver cells, whether temporary or chronic, is a likely initial cause of the insulin resistance we've been talking about and that we're trying to prevent and/or reverse. In short, insulin resistance starts in the liver and then becomes systemic.

All this science is still speculative, as is the contention that sugar is uniquely addictive (although if you have either children or a sweet tooth, you likely don't need a lot of science to accept it). When adolescents with fatty liver disease stop consuming added sugar (as in a trial funded by my nonprofit organization, the Nutrition Science Initiative, and published in the medical journal **JAMA** in January 2019), the fat in the liver tends to go away. This suggests that insulin resistance—in children, at least—would resolve along with it.

All the other carbohydrates in our diet—glucose, lactose (in milk), maltose (in beer), and others—work more or less directly to make us store fat by raising blood sugar and so stimulating insulin. Sugar does it both directly and indirectly: The glucose raises blood sugar and stimulates insulin secretion, the fructose overwhelms the liver and causes fatty liver and insulin resistance, so that we secrete ever more insulin to all those other carbohydrates.

Brillat-Savarin's observation that sugar makes

everything worse when it comes to getting fatter still holds. If there's a primary evil in this nutrition story, it's almost assuredly sugar, and learning to avoid it and still enjoy both life and eating is key. It may not return you to health and correct your weight; that's likely to require Brillat-Savarin's more or less rigid abstinence as well. But it is step one in preventing the problem from getting worse.

While I'm making the case for abstinence, it's important to realize that it is not a panacea. It does not mean that anyone who is obese will become lean, only that they will very likely become leaner and healthier, and they will do so without hunger. Other hormones influence fat accumulation, sex hormones in particular, and they do not respond directly to what we eat (although they may indirectly). Insulin is the direct primary connection to our food. For many of us, we will have to minimize our insulin secretion to create and prolong that negative stimulus of insulin deficiency, to mobilize and burn more fat than we store, to achieve and maintain a healthy weight. More or less rigid abstinence will indeed be both necessary and ideal.

Ultimately, your success will depend on your commitment. While this may be said for every diet, the commitment here is not to living with hunger. Some who need to lose a dozen pounds to attain what they perceive as their ideal weight and health might do fine just by cutting back on the more obviously fattening foods and the carbohydrates they contain—for instance, sugary beverages, beer ("shun beer as if it were the plague," wrote Brillat-Savarin), desserts, and sweet snacks. These folks will do fine eating slow carbs, with their complement of fiber to slow digestion and absorption and keep insulin levels low. Rigid abstinence will not be necessary **for them.**

For most of us who have struggled with our weight for years or decades, however, rigid abstinence would be ideal. Physicians who recommend LCHF/ketogenic eating say they will settle for the best their patients can do, but they believe that the benchmark for how healthy we can be comes only with rigid abstinence. The physicians who have worked with obese patients the longest and whose clinics have accumulated the most experience, like Eric Westman at Duke University, are adamant. "The word on the street," says Westman, "is that I'm too strict. But maybe you have to be strict."

At a recent diabetes conference in Aspen, Colorado, I had the opportunity to speak with a young woman who had participated in a diet trial that my nonprofit had supported at Stanford. She had been obese her whole life, she told me, and weighed 240 pounds at the start of the trial. She was randomized to be among those participants who would follow an LCHF/ketogenic eating plan for a year. For the first three months she practiced rigid abstinence and lost thirty pounds without the obsessive thoughts of food and the hunger that accompany calorie-restricted diets. (She had charted her weight on an app on her smartphone, and this is what she showed me.)

Then the Stanford researchers suggested that she and her fellow study participants who were assigned to LCHF/ketogenic eating could and maybe should go back to eating small portions of carbohydrate-rich foods that they specifically missed. The researchers were worried that if the diet was too restrictive, the subjects might fail to sustain it and would drop out of the trial. So this young woman went back to eating berries, which for many of us would be benign, but now she lost only five pounds over the next three months. At the six-month mark, again on the advice of the researchers, she added back a little more fruit and never lost another pound.

It's certainly possible that her weight would have plateaued even without the berries and then the fruit; we can never know. But neither will she— and that's the point. Had this young woman continued with rigid abstinence, she might have lost significantly more weight. If so, she might have decided that rigid abstinence was clearly worth the effort, and that a berry- and fruit-free life was eminently worth living. As self-help and management advice books will often say, setting a goal and committing to it are vitally important. Without the commitment, we never get to find out if the goal is achievable. By diluting the commitment and allowing us to compromise, we never know.

To abstain more or less rigidly from sugary, starchy, and floury foods means we have to change the way we think about how we eat, the foods we eat and don't eat, and the effort we put into thinking about every meal. Like anything that requires discipline, however, it gets easier the longer we do it. In this case, we have an advantage over other similar lifelong interventions: By changing what we eat, we're changing our physiology, the very fuel that our cells need to survive and generate energy, and that in turn should change the type of foods for which we hunger. As our bodies learn to burn fat exclusively for fuel, it's fat we should begin to crave—the butter rather than the toast.

Temptations will never vanish. A sugary treat may not be any less seductive than it ever was. Sugar may always have the power to excite our taste buds (and our liver) and trigger cravings for more. But the key is to not succumb. As our bodies switch to burning fat for fuel, the ability to say no to sugary treats will be reinforced. Many foods with sugar in them will taste too sweet as our tastes change. This is commonly reported. We will also become more adept at and habituated to keeping our lives and our environments sufficiently sugar-free and so temptation-free. Successfully quitting any addiction means learning to make the source of the addiction unavailable, whenever possible. Success will require making a commitment to an objective and then being both patient and resolute in achieving it and maintaining it.

Many of the physicians I interviewed for this book spoke about their own health and approach to LCHF/ketogenic eating in addiction terms. Robert Cywes, a pediatric surgeon who now runs bariatric surgery and weight-control programs for adults and adolescents in Florida, said to me, "To cut to the chase, we are a carbohydrate substance abuse program, not a weight loss program." Martin Andreae, a general practitioner in Powell River, Canada, just north of Vancouver, described himself as a reformed sugar addict.

"One brownie and I should be done," Andreae said. "My common sense says stop there, but my actions don't. I understand the feeling of addiction, the powerlessness of it. But the joy we get from an addiction is filling the void created by the absence of the substance itself. And you don't cure an addiction with moderation; you do it with abstinence. Any other addiction field, that's how we treat it. Alcohol: We say stop it altogether and don't even have alcohol in the house. It's the same with smoking. With diabetes and obesity, your body is essentially a sugar or carbohydrate addict. Telling our patients to moderate intake is telling them to do something that is almost physiologically impossible and keeps the addiction alive. What we're fighting against is the concept of moderation. It doesn't work."

Mark Cucuzzella, a physician and professor of medicine at the West Virginia University School of Medicine, referred to himself in our interview as a "prediabetic in remission," while implying that he is a carbohydrate addict in recovery. Cucuzzella is a marathoner, author of a book on running and health (**Run for Your Life**), and he eats and prescribes LCHF/ketogenic eating. His conversion experience was prompted by a diagnosis of prediabetes despite weighing only 135 pounds (he's five foot eight) and religiously running ten miles a day. He says he was "literally" eating carbohydrates every three to four hours, including at two in the morning. He described his life, day and night, as "hungry, eat, hungry, eat, hungry eat. . . . My last bowl of cereal and my last piece of bread were over six years ago. I do not miss them."

Using language from Gretchen Rubin's **The Happiness Project,** Cucuzzella divides his patients into "moderators" and "abstainers." "A moderator can eat one little square of dark chocolate and walk away," he says. "An abstainer has one bite, and it will not go well—he'll eat the whole damn bar. One of the messages that has been a complete disaster for patients with obesity and diabetes is that we can do this in moderation. But if you're really carb-addicted, telling you to go from ten doughnuts down to four is just telling you to think about eating the doughnuts all day. A rare patient can be a moderator when it comes to tasty carbs and succeed. Most of us need to be abstainers. Like people with alcoholism, drug addiction, and smoking, we need to avoid completely, and then we'll have better odds of success. Why this advice is considered 'extreme' is beyond my comprehension as I witness patients daily who suffer from these metabolic diseases."

14

Defining Abstinence

Abstinence from everything that is starchy or floury and sugary means: Don't eat those foods.

I'm often reminded in my conversations about diet and health that what one person thinks about for twenty years (obsessively) is not necessarily obvious to those who are thinking about it for the first time. So let's return to basics: what we're not going to eat, what we're abstaining from eating, and what we can eat freely.

Abstaining from carbohydrates and carbohydrate-rich foods means you won't be eating the foods in the list below. You won't be eating them because they are predominantly carbohydrate and so will raise your blood sugar, stimulate insulin, and promote fat accumulation and hunger.

- No grains, which means no rice, wheat, corn, or even "old world" grains like

quinoa, millet, barley, and buckwheat.
No products made from these grains:
no pasta, breads, bagels, cereals. No sauces
that use cornstarch as a thickening agent,
as many do.

- No starchy vegetables, so no root vegetables
or tubers. No potatoes, sweet potatoes,
parsnips, or carrots. You won't eat vegetables
that grow below ground. It's okay to eat
those that grow aboveground.
- No fruit, with the exception of avocados,
olives, and tomatoes (all technically fruit),
and with the possible exception of berries,
which we'll discuss.
- No beans or legumes, which means no peas,
lentils, chickpeas, or soybeans.
- Absolutely no sugary foods and particularly
sugary beverages, even if the sugar comes
from "natural" sources like fruit: so no soda,
fruit juice, smoothies, cakes, ice cream,
candy, bonbons, or even health-food bars,
and perhaps particularly those advertised as
low in fat.
- No milk or sweetened yogurts, particularly
low-fat varieties (in which the fat content is
removed and replaced, typically, with some
kind of sugar). I agree with Michael Pollan
that if a food product makes a health claim
on its packaging, it's probably a good idea to
avoid it.

In general, the more fiber a food or food product contains and the greater the proportion of calories from fat, the lower the blood sugar response, the lower the insulin response, and the more benign this food might be. Research suggests that we have huge individual variation in how our blood sugar responds to different foods, which implies a huge variation in insulin as well. Maybe potatoes are benign for some of us but not others. The problem is we don't know, and if we did, "more benign" might not be good enough. So the best advice, if we're committing to being healthy and ideally lean, is to abstain from all.

Below are the foods that you can eat, the foods that are very low in carbohydrates and/or high in fats.

- Meat: from animals or fowl (chicken, turkey, duck, goose), the fatter the better and all preferably raised on grass, in pastures, and not in factory farming conditions
- Fish and shellfish
- Eggs

You can cook these foods any way you like (baking, broiling, stir-frying, roasting), but you have to avoid using flour, breading, or cornmeal in the preparations. You can also eat:

- Butter, preferably from grass-fed animals, and oils, preferably from fruits rather

than nuts, seeds, or legumes, and so olive, coconut, or avocado oil
- Low-carbohydrate vegetables, which means all leafy green vegetables, in particular, kale, spinach, and lettuce, but also cabbage, broccoli, cauliflower, asparagus, Brussels sprouts, tomatoes (technically, a fruit, as mentioned), mushrooms, cucumbers, zucchini, peppers, and onions
- Fatty fruits: olives and avocados
- Dairy fats: cheeses, cream, (unsweetened) yogurts, all full fat

These foods you can eat but in moderation, as I'll discuss.

- Low-sugar chocolates, the lower the better
- Berries
- Nuts and nut butters
- Seeds and seed butters

I said "in moderation" for these latter foods because they fall on the borderline of acceptability: Clinical experience suggests that they can be a problem. Once again, individual variation plays a role in how our bodies tolerate these foods. Fat constitutes the bulk of the calories in nuts and seeds and their butters, which is a good thing, but they can still have sufficient carbohydrates to stimulate insulin and so fat accumulation and a craving to eat more, a bad thing. The better the nuts taste, the greater the carbohydrate content tends to be. Most lists of foods permissible in ketogenic diets include nuts and seeds and butters made from them. You can now buy flour made from nuts and seeds and bake with it. You can buy granolas made predominantly from these foods and have them for breakfast. Snack bars, too, of course. Most of the physicians I interviewed think of nuts and seeds and their butters as necessary fat-rich snack foods in LCHF/ketogenic eating. That's the general consensus, but . . .

The allowance of nuts, seeds, and their butters and the issue of individual variation come with an obvious warning: If you're not losing your excess weight while otherwise embracing LCHF/ketogenic eating, then these foods may be a problem **for you,** and you should see what happens when you abstain. Eric Westman is, once again, strict on this account: Nuts and nut butters, seeds and seed butters, are not included among the foods he counsels his patients to eat. His experience tells him that his patients too easily overconsume these foods. They think they're eating a modest amount, and they're not. They eat them even when they're not hungry.

With the exception of olives and avocados, in which the calories come predominantly from fats, berries fall on the borderline of acceptability, while large fruits—apples, pears, oranges, grapefruits, pineapples, and melons—should be avoided. The carbohydrates in these fruits are less concentrated than they are in starches because of the water content of the fruit. But they still generate a blood sugar and insulin response and are still likely to be fattening. An apple is sweet to the taste precisely because it contains both fructose and sucrose. They are bound up with fiber and so are far slower to digest than they would be in soda or fruit juice. The lean of the world might tolerate them effortlessly. The rest of us probably can't.

Berries, though, are relatively low in carbohydrates and sugar and high in fiber and perhaps low enough to be acceptable. But there's a catch even here: While berries were available to our ancestors, who probably consumed them over the course of the last million or so years, they would have done so only seasonally, a few months a year, and they would have tended to eat them on the sour side, before they fully ripened. Even fully ripe, these berries would likely have been less sweet than the varieties available today at the market.

Where I live, in Northern California, blueberries are in season for about six weeks a year. They appear in bins in my local market and are indescribably (for me) delicious when they do. I eat them in immoderate quantities. It's quite likely that I fatten a bit during that period, but then they pass out of season (as the growing season moves north), and I no longer buy them and, I hope, lose whatever weight I gained. (And then blackberries come into season. . . .) Eat them all year round, and there are no guarantees.

15

Making Adjustments

Abstaining from carbohydrates does not imply eating less; it implies eating fat and fat-rich foods.

What does it mean for a weight-control diet to be meaningfully sustainable? Health journalists and nutritional authorities will now insist that the best diet—the one that they say "works"—is the diet that we can sustain, to which we can stick for life. But what does that mean? Sustaining a diet that doesn't help us reach and maintain a healthy weight is of little benefit and clearly isn't one that's working. And to sustain a way of eating for life, almost by definition, we have to be able to eat to satiety. That implies we're not walking away from our meals hungry. It implies we're not counting our calories; we're just eating, as lean people do. Anything that requires a lifetime of hunger (in a world in which food is abundant) is a promise of failure.

This is why outside the world of academic research, in the fad diet world and that of the physicians with hands-on experience, LCHF/ketogenic eating prescriptions come **without** the advice to count calories or eat less. The technical term is **ad libitum:** Eat as much as you like. Eat when hungry, and eat until satiation. Physicians who advocate this way of eating to their patients, particularly those with the most clinical experience, tend to be adamant that their patients eat whenever they're hungry. The expectation is that if we don't, we will eventually give up on the diet, or we will binge-eat in response to the deprivation, losing the health benefits.

To make this work in practice, to abstain from carbohydrate-rich foods while eating to satiety, we have to eat significant amounts of fat. Carbohydrates typically constitute half of the calories we consume. So if we're abstaining from carbohydrate-rich foods and the energy they supply, then we're going to have to replace some large proportion of those calories by eating more protein or more fat, and real food sources of protein invariably come with significant fat attached.

While nutritionally adequate eating requires a minimum amount of protein for lean tissue repair and growth, the protein itself is composed of amino acids, and these can be converted to glucose in the liver and then stimulate insulin secretion. This is a slower process than eating refined grains or drinking sugary liquids, but the result is still likely to be at least some insulin secretion. If your fat cells are exquisitely sensitive to insulin, even this amount might be too much. An eating pattern that **minimizes** insulin is not high-protein. This would have been less of a problem in the 1960s, when the typical meat sold in supermarkets and by butchers was 70 percent fat by calories and people ate their poultry with the skin attached. But as the anti-fat message was broadcast widely and we turned to leaner cuts of meat (like the skinless chicken breast) and lean fish, eating to avoid carbohydrate-rich foods can all too easily mean eating too much protein.

Say you eat for lunch or dinner a skinless chicken breast and green vegetables or a green salad. This kind of meal seems like an eminently reasonable compromise between nutritional paradigms. It has no starchy vegetables, grains, or sugars and so is low-carb and can seem suitable for ketogenic eating. The skinless chicken breast keeps it low in fat as well. It's easy to understand, in the midst of this endless nutrition controversy, why we might want to hedge our bets this way. Maybe people like me are right in arguing that the major problems with modern diets are the refined grains and sugars, but it's also hard to believe that the anti-fat authorities got it **all** wrong. Thus what seems like a happy compromise: restricting your fat consumption, while getting your carbohydrates from sources we

all agree are benign—specifically, nonstarchy vegetables. The skinless chicken breast has plenty of
protein and not that much fat. The carbohydrates
in the meal are "good" carbohydrates, "slow"
carbs. They're bound up with fiber, and we'll digest them slowly.

But the devil, as ever, is in the details. If the portions are small enough and if we eat slowly enough,
the insulin secretion from the amino acids in the
protein and from the carbohydrates in the green
vegetables may indeed be insufficient to rise above
the insulin threshold. We'll still be burning more
fat than we're eating. It doesn't flip the switch on
that insulin threshold. But not so for large portions, obviously. If we're consciously choosing
small portions, we'll likely be hungry afterward. If
we're hungry, we're likely to cheat on our diet, or
quit it entirely. We might be able to eat like this
while we're losing weight, because we'll be burning
our own fat, too, but what about once we plateau
at a healthy weight? If we increase the portion size,
though, the insulin response increases as well. Eat
enough calories to be satiated, and it can reasonably be expected that we will store more fat and be
hungry between meals, while still hungering for
carbohydrates. It can be a recipe for eventual failure, reasonable as it may seem.

The only way to eat a satiating meal while minimizing insulin secretion is to add fat. It's the one
macronutrient that does not stimulate an insulin

response. When Australian researchers led by Jennie Brand-Miller of the University of Sydney studied the effects of mixed meals on insulin secretion—the only ones to publish such a comprehensive study (2009), as I write this—the best predictor of insulin secretion was the fat content. The higher the fat content, the **lower** the insulin response. "Because protein stimulates insulin secretion, particularly when combined with carbohydrate," they wrote, "the meals with the highest protein and carbohydrate content (and hence lowest fat content) produce the highest insulin responses."

What about meals that are high in both fat and protein? I've heard from readers over the years who have taken the guidance from my books and others and applied it by eating three meals a day of fatty meat—rib eye steaks for breakfast, lunch, and dinner. In the growing world of people who describe themselves as carnivores or "zero carbers," they don't even eat green vegetables. Steve Phinney and Jeff Volek, who have done more research on ketogenic diets than anyone, believe that there's an upper limit to the amount of protein we can eat and remain in ketosis—less than a gram of protein per pound of body weight.

Whether that much protein would inhibit mobilization of fat from the fat cells and ultimately shorten our lives is the kind of question that remains unanswered. The young man I mentioned earlier who weighed close to 400 pounds when he was eighteen preceded to lose over 120 pounds in four months of eating nothing but fatty meat that his father bought for him by the tens of pounds a week at Costco. His response to this kind of diet may be relatively rare if not freakish, but it could also be the norm. And even my friend's response might change with time and age. Maybe this is the response of an eighteen-year-old male predisposed to obesity but not that of a forty-year-old or even an eighteen-year-old female. We have no way at present of telling.

The huge amount of individual variation in how our bodies process both protein and carbohydrates means you will have to experiment and find what works for you. No meaningful clinical trials have been done comparing LCHF/keto eating to what we might call LCHP—low-carbohydrate, high-protein. As discussed, consuming protein will also stimulate secretion of two hormones—glucagon and growth hormone—that work to get fat out of fat cells. These diet-induced hormonal responses are less well studied than that of insulin. What's lost with protein because of the insulin secreted might be gained back by the glucagon and growth hormone response. Even if it is, though, and our meals are particularly rich in protein, replacing the carbohydrate calories we're not eating will still require plenty of fat and fat-rich foods.

This is why the Indiana University physician Sarah Hallberg tells her patients that green vegetables are a conduit for fat and should never be consumed without it. Hallberg is the medical director of Virta Health and oversaw the start-up's LCHF/ketogenic eating trial on patients with type 2 diabetes. The subjects with diabetes in the Virta Health trial got the same advice: When cooking vegetables, do so with copious butter or olive oil, then eat them with olive oil or melted butter. Lunch can be a salad so long as the salad dressing has plenty of fat and is low in carbs. Put olives or avocado on the salad or perhaps hemp seeds. A good salad dressing, says Hallberg, has plenty of oil and fewer than two grams of carbohydrates per serving. She recommends taking such a salad dressing, splitting it into two containers, adding more olive oil to each, and shaking them up to increase the fat content. By using vegetables as a conduit for fats, LCHF/ketogenic eating can be mostly plants if not all plant foods. It may be harder to do without the fatty animal products, but it's certainly doable.

What is it like to eat to satiety on foods that are very low in carbohydrates but high in fat? Is this way of eating as radical as it's often portrayed? I'm going to use pictures to answer these questions. In the process, I'm going to take the opportunity to demonstrate why weight control is less about how much we eat and far more about what we

Two versions of a dinner just over six hundred calories. The fattening meal (top): a roasted chicken breast, broccoli, and potatoes.

The non-fattening/weight-loss meal (bottom): two chicken thighs, more broccoli, no potatoes, butter.

eat. That's a primary reason it's so helpful to stop thinking about how many calories you're eating and how much you're burning off in exercise. It confuses the matter; it doesn't clarify, not if you want to achieve and maintain a healthy weight.

What follows is a day's worth of meals in pictures, beginning with dinner and working backward to breakfast. The dinner plate at the top—a roast chicken breast, broccoli, and potatoes—is fattening to those of us who are predisposed because of the carbohydrates in the potatoes. The plate at the bottom—roast chicken thighs for the higher fat content of their meat, and broccoli with butter (or olive oil per Hallberg's guidance)—is not. It's part of a weight loss and maintenance way of eating. The two plates of food as pictured contain essentially identical calories—just over six hundred. One has potatoes and is fattening; the other one doesn't, has a larger portion of chicken (by calories because of the greater fat content), more broccoli, and the butter on the broccoli. It's not fattening. The larger portions of chicken and broccoli and the butter (or olive oil) make up the difference in calories. If you were ordering this in a restaurant, you would order the roast chicken and ask the waiter or waitress to replace the potatoes with more broccoli or a green salad. Simple enough.

David Unwin, a general practitioner in England who in 2016 won the National Health Service innovators award for advocating LCHF/ketogenic eating to his patients with diabetes, describes this as "turning everything that was white on your plate to green." Even with equal or greater calories, the plate on the bottom is part of a weight-loss program (a fad diet, Atkins!); the plate on the top is likely what you've been eating all along and has contributed to making you fatter.

Eating dinners like the one on the bottom should be easy to sustain. All you're doing differently is **not eating** a potato and eating your vegetable with butter or olive oil. As for heart health, virtually all authorities would consider the meal on the bottom to be as healthy as the one on the top, certainly if the added calories come from olive oil. So that would be the compromise. If you choose butter instead of the olive oil, you're assuming that all I've told you in this book is correct.

Lunch could be identical to dinner, with the same implications about sustainability and health, but let's give it a fast-food, standard-American-diet twist. The plate on top, unappetizing as it may appear, is a typical fast-food meal: a McDonald's cheeseburger on a bun (along with pickles, onion, ketchup, and mustard), a small order of french fries, and a small Coca-Cola. It has about seven hundred calories (with the ketchup) and is fattening to those predisposed because of the bun, the fries, the sugar in the soda, and even the sugar and carbs in the ketchup. The plate at the bottom has a Double Quarter Pounder with cheese (along with lettuce, tomatoes, onions, and pickles), no bun, salad and ranch dressing, no fries, and water instead of soda. It has the same number of calories but without the grains (the bun), the starches (the fries), and the sugar in the soda and ketchup. It's not fattening. The two meals have

Two versions of a seven-hundred-calorie lunch. The fattening meal (top): A small cheeseburger, fries, ketchup, and a small Coca-Cola.

The non-fattening/weight-loss meal (bottom): a Double Quarter pounder with cheese (no bun), a green salad with ranch dressing, and ice water.

equivalent calories but different carbohydrate content, and they create different metabolic, hormonal responses—different effects on fat accumulation.

The fast-food meal at the top makes you fatter. The fast-food meal at the bottom makes you leaner. It fits into LCHF/ketogenic eating. It's the rare health expert today who would suggest that a meal with two hamburger patties instead of one, plus a salad, is less healthy than a meal with the one burger plus fries and a sugary beverage. If you showed the health experts only the picture at the bottom, they might mutter about the red meat, but they'd probably accept that it's healthy even by their predilections, so long as you weren't eating "too much." If you replaced the two hamburgers with a nice piece of salmon or salmon burgers (still no bun) or even an Impossible Burger (meat-free, no bun), we would mostly all be in agreement: a healthy meal.

Breakfast seems to be the ultimate battleground, the meal that diverges most radically from conventional healthy thinking. This is the bacon-and-eggs problem. The authorities for the last fifty years did a very effective job in convincing us these were agents of death. We came to believe that the just-over-seven-hundred-calorie breakfast at the top of the following page—cereal, skim milk, banana slices, toast (buttered), and juice—is ideal, yet that breakfast is fattening to those predisposed because of the carbohydrates in all those (including the lactose in the milk). Because of its effect on blood sugar and insulin, it will leave those of us who are insulin resistant and predisposed to fatten likely to be hungry later. We'll want a midmorning snack, likely a carbohydrate-rich one. The plate at the bottom—three eggs scrambled with cheese and sausage, two strips of bacon, avocado slices, and water instead of juice—has the same number of calories (approximately seven hundred) and is not fattening to us. And because insulin remains low, we won't be hungry later; we'll have no urge for a snack.

The three plates at the top constitute the standard American diet. With the exception of the fast-food lunch, the nutritional authorities would consider them part of a healthy lifestyle. But they are what most of us have been getting fat on, along with between-meal snacks of much the same macronutrient composition, and then the sugar-rich or carb-rich beverages, sodas, beers, and so on. The three plates at the bottom have the identical calories and are part of a weight-loss diet, an LCHF/ketogenic eating pattern—i.e., Atkins or keto—that will allow you to achieve and maintain a healthy weight.

It's not about the calories they contain. While some might look at the LCHF/ketogenic lunch plate at the bottom and say they can't eat that much food for lunch (or at least not without a significant number of those calories coming as the sugar in

Two versions of a breakfast of just over seven hundred calories. The fattening meal (top): cereal, half a banana, skim milk for the cereal (four ounces), buttered toast, and orange juice (eight ounces).

The non-fattening/weight-loss meal (bottom): three eggs scrambled with cheese and sausage, two strips of bacon, half an avocado (sliced), and ice water.

the soda), others can imagine it effortlessly. They would very likely still lose weight or maintain a healthy weight eating it, because the fattening comes with the carbohydrates, not the calories.

These pictures also inform our understanding of sustainability, which is required for any dietary intervention to succeed. It's true that the LCHF/ketogenic lunch requires a fork and maybe even a knife, and it certainly can't be consumed while driving without creating a mess, which isn't necessarily the case with the standard-American-diet version. But otherwise, what you're primarily doing when you eat LCHF/ketogenic food is **not** eating certain foods, and so sustainability is about whether you can keep that up. When cigarette smokers quit smoking, the quitting is sustainable only so long as they don't smoke cigarettes. The same logic holds for LCHF/ketogenic eaters and their abstinence from carbohydrate-rich foods.

Does eating more fat to compensate for the carbohydrate calories make it unhealthy? In the 1960s and '70s, the British nutritionist John Yudkin pointed out that when we restrict carbohydrate-rich foods—specifically, grains, starches, and sugars—we are restricting the foods that bring the least to the diet in terms of vitamins and minerals. In the case of sugar, it brings nothing at all but energy (hence the term "empty calories") and a metabolic burden to the liver that may very likely be the cause of insulin resistance.

The science of metabolic syndrome and its link to obesity, diabetes, and heart disease, as we discussed, implies that the carbohydrate-rich foods we have to avoid to attain a healthy weight are the same as those we have to avoid to attain and maintain good health. The evidence implicating natural dietary fats in heart disease has evaporated over the years. Because the LCHF/ketogenic meals eaten at the bottom in the photos will help us achieve and maintain a healthy weight, they are also correcting metabolic syndrome. We have significant evidence now that they will even reverse type 2 diabetes. These foods, including the fat, are integral parts of a healthy diet.

Another principle that we have to accept then is that these naturally occurring fats can both be good for us and constitute the great majority of the calories we consume. These are the fats from animal products—whether saturated or not, even lard and tallow and chicken fat—and the fats from vegetables that include oils we've been consuming for thousands of years, olive oil in particular, and oil from avocados. We've been eating these fats long enough as a species that we can consider them natural, as Geoffrey Rose might have defined it, and so believe with reasonable certainty (the best it gets) that these foods are benign. Will we shorten our lives by eating so much fat or red meat? The existing clinical trial research suggests that the answer is no, though there are no guarantees. The simple fact, though, is that in the short run, we get healthier.

16

Lessons to Eat By

You don't get cake and ice cream when you're finished.

The advantage of covering the field of obesity, nutrition, and chronic disease as a journalist rather than as a physician or a researcher (or a blogger), as I noted earlier, is that the job ultimately is to learn from people who have first-hand observations of the subject in question. The more such people you speak with, the more you learn. As I also noted earlier, I spent half a year interviewing physicians who now prescribe LCHF/ketogenic eating in their clinics and dietitians who prescribe them to their clients and eat this way themselves, as well as a few dozen other health care practitioners.

Among the physicians were those who struggled to get this dietary message across in the fifteen minutes that their health care system allowed them to allot to each patient, and others who had trans-formed their entire practices to focus on weight control and the prescription of LCHF/ketogenic eating and had hired only nurses, dietitians, and staff physicians who bought into this paradigm as well. Charles Cavo, for instance, began his medical career as an obstetrician/gynecologist working in central Connecticut. In 2012 he decided that he would be failing in his job if he did not also provide his patients with counseling for obesity and diabetes. When his partners "weren't interested and thought [he] was crazy," he started his obesity medicine practice on the side: giving advice, he told me, to "two people in his kitchen." He has now seen and prescribed LCHF/ketogenic eating to over fifteen thousand patients and had to leave his OB/GYN partnership to keep up with the obesity practice. Two of the physicians I interviewed—Sean Bourke in Northern California and Garry Kim in Southern California—had established chains of weight-loss/weight-management clinics that originally counseled clients to use traditional calorie-restricted diets, even very-low-calorie diets, for weight loss and then evolved over time to prescribing LCHF/ketogenic eating instead.

Bourke, a Yale-educated emergency medicine physician, is the cofounder of the dozen JumpstartMD clinics in the San Francisco Bay Area. He told me that some fifty thousand patients had come to these clinics looking for advice on controlling their weight since he opened the first one in January 2007. This is, in effect, his clinical experience. (With his

JumpstartMD colleagues and a collaborator at the Lawrence Berkeley National Laboratory, Bourke recently published a paper in the **Journal of Obesity** on the results from over 24,000 of these patients, for whom he had complete clinical data.) The program, he said, originally counseled a broadly calorie-restricted approach—"low in everything"—and its patients achieved what Bourke called reasonably good results so far as weight loss was concerned. But the patients were also, not surprisingly, always hungry, and they would have to deal with that hunger forever, if they wanted to maintain that weight loss. "We were seeing better results in the lower-carb, higher-fat state," Bourke said. "We were seeing people who were just less miserable and less dependent on medications to suppress their appetites. If they embraced the low-carbohydrate, high-fat, they found it a broadly more sustainable way to eat, with a better flavor profile, greater satiety, and greater craving reduction. Over time medications were not as necessary and the lifestyle feels more sustainable to them, if they embraced it."

That last clause, which Bourke repeated twice, has always been the critical one in any dietary program— **if they** (i.e., you or me) **embrace it, it will work.** I've tried to provide the rationale, biological and historical, for why LCHF/ketogenic eating is worth the effort, but you have to make that effort. Believing in what you're doing and doing it for the right reasons are both essential conditions for success.

Before giving the simple practical advice I gathered in the course of my interviews, I want to share six lessons that capture the essence of the practice of LCHF/ketogenic eating: what we're trying to achieve and how to go about doing it. In short, I want to suggest how you should think about how to eat in order to achieve and maintain health and a healthy weight. Five of the lessons are from the practitioners I interviewed for this book, but the first is from Michael Pollan and his 2008 best seller **In Defense of Food.**

Much of this book has been a repudiation, mostly implicit but not always, of the relevance, to those of us who are not naturally lean and healthy, of Pollan's otherwise seemingly sensible mantra—"Eat food. Not too much. Mostly plants." For us, "not too much" is meaningless. "Mostly plants" is not ideal and may be to our detriment (ideal as it may be for the animals and maybe even the environment, although that, too, is not as simple as it is often portrayed).

Even the advice to "eat food" rather than foodlike substances is something on which I often find myself bristling. Not because I don't believe eating whole foods is an essential component of healthy eating, which I do.* But it's far from sufficient to

---

* I get nervous when I see keto- or paleo-friendly highly processed foodlike substances now appearing in markets and online, sweetened with coconut sugar or noncaloric sweeteners or anything else. They may be benign, but they also may not be.

imply that if a food is not processed it's benign, at least for those who are predisposed to put on fat. As clever as this wording is (and as much as I often invoke Pollan's "foodlike substance" terminology myself when discussing/arguing with my family about the pros and cons of snack foods), the implication is that this is enough to guide those who are not lean and healthy back to health. It's not. " 'Just eat real food' is perfectly wonderful advice for preventive medicine," as the San Antonio physician Jennifer Hendrix, founder of the Women Physicians Weigh In Facebook group (with over thirteen thousand members as of fall 2019), said to me. "But once a person has obesity, and particularly obesity with comorbidities like diabetes and hypertension, it's much more complicated. I've never seen anyone who has had a weight problem their whole life have the weight go away merely by changing to a real food diet, because some of those real foods are still fattening."

1. "Many of the policies will also strike you as involving more work. . . . In order to eat well we have to invest more time, effort, and resources in providing for our sustenance, to dust off a word, than most of us do today."

Michael Pollan and I have disagreed on many points, but not on this message from **In Defense**

of Food. Getting healthy and staying healthy, regardless of weight, involves work and a lifetime commitment. This is true of all thoughtful advice on diet and health. How we eat plays a critical role. As informed physical trainers will tell their clients, you cannot outrun a bad diet. Eating healthy requires thought, planning, and more work than simply reaching for the default choices easily available in our daily lives. As Pollan implies, it's certainly not the kind of thing that we're typically trying to do when we're pursuing food that's cheap, fast, and easy, as the standard American diet typifies. Even then, though, it's not impossible. It just requires greater effort.

The practitioners I interviewed were unanimous in their belief that LCHF/ketogenic eating will lead to improved health and substantial weight loss without hunger for all but the rare exceptions. But we have to be willing to embrace it, to make the effort to commit to eating as prescribed. Those who succeed are those who come to think of abstaining from carbohydrate-rich foods as critically important to their health, in the same way that ex-smokers consider continued abstinence from cigarettes vitally important and members of Alcoholics Anonymous consider the act of abstaining from drinking. This means you will have to figure out how to avoid temptation in a world that will serve it up, literally, by the platefuls. "We live in a carb-centric world," as Kathleen Lopez, a dietitian who teaches at Illinois's Dominican University, said to me. "Everywhere you go, everyone is eating ice cream and potato chips, and you're there, and you're not. For some people that is not difficult at all; it's an even trade-off for their improved health. For other people it's torturous."

For those who do embrace this new way of eating, some will get healthier and leaner than others, but all should get healthier and eventually find it easy to do. This is true of the process of breaking any addiction. In this case, the replacement of carbohydrate-rich foods with fat-rich ones should provide the pleasure and joy in eating that might not come from the kind of low-fat, calorie-restricted meals we've been told since the 1970s that we have to eat to avoid heart disease and remain lean. It does not require a lifetime of hunger, only a lifetime of abstinence from a specific food group that for us (but not everyone) is harmful.

2. "This is not something you are going to do. This is what you are going to become."

This is what Ken Berry, a physician with a practice in rural Tennessee, tells his patients when they come to him looking for advice. Berry has been practicing since 2003 and began prescribing LCHF/ketogenic eating to his obese, insulin-resistant, and diabetic patients half a dozen years later. (He's also the author of a 2017 book, **Lies My Doctor Told Me,** largely about the benefits of LCHF/ketogenic eating, specifically compared to the low-fat, calorie-restricted advice our doctors have been traditionally passing along.) By Berry's advice that "this is what you are going to become," he means his patients are going to become people who are as meticulous about what they eat and how they eat as they might be in any other area of their professional or personal life.

Berry's conversion experience is typical of many of the physicians I interviewed. He started putting on excess fat in his mid-thirties and believed (based on what he described as the "four hours of nutrition education we got in medical school") that "if you want to lose weight you create a calorie deficit." He assumed that if he religiously followed this "state-of-the-art" wisdom, it would fix him. It didn't. "So now I looked like a fat lazy doctor, with stiff joints and reflux and allergies," he told me. "I felt like shit all the time, and I'm supposed to go into a patient's room with my gut hanging over my pants and tell him what to do. How does that work?"

Instead he read **The South Beach Diet,** which took him to Atkins, and Loren Cordain's book on paleo eating, and then **The Primal Blueprint** by Mark Sisson. Finally he started looking into ketogenic diets and the benefits of intermittent fasting. Along the way he changed what he ate and saw "immediate results." Even his allergies and reflux went away, he said, which he attributes specifically to giving up dairy. "I came to the conclusion that everything I was telling my patients to do and not to do was exactly wrong. This is not a fun or comfortable realization as a professional who's supposed to know what the hell he's talking about."

Now Berry freely acknowledges to his patients that he "sounds like a witch doctor with a panacea" when he counsels them how to eat, but he believes, based on his clinical experience, that LCHF/ketogenic eating resolves a host of conditions, including obesity, diabetes, and hypertension. The key, of course, is getting his patients to believe it as well.

Berry is among the many physicians and dietitians I interviewed who stressed the importance of patients "going down the rabbit hole" or "doing the homework," as he did, that they make the effort to learn why they should eat as he advises and what to expect. The patients who succeed, these physicians agreed, tended to be those who could be induced to read at least some of the copious literature on LCHF/ketogenic eating that's now available. They had to become people who cared enough to do the work. "I'm pointing out to them websites, pointing out books to read," Berry says. "And if somebody is not willing to do that, if they're not ready to do their homework, they're probably not ready to change. That's fine, if that's the case, even if they're becoming diabetic, and I'll tell them so. But I'll also tell them that I will see them in a couple of years when they're getting started on insulin for their diabetes and might be a little more motivated."

3. "You don't get cake and ice cream when you're finished."

Nick Miller, a dentist who has a practice in the Pittsburgh suburbs, said this to me. It's what Miller tells his patients in conveying the idea that this is a lifetime commitment.

Miller's conversion experience was another common one. He graduated from high school at 190 pounds, a six-foot-two athlete. Eight years out of dental school, his weight had ballooned to 280 pounds. It happened, he said, despite his running fifty-plus miles each week "and trying to avoid bad foods." Miller's trip down the rabbit hole began with a podcast, in which Vinnie Tortorich, a fitness trainer in Los Angeles and now an author and documentary filmmaker, was discussing LCHF/ketogenic eating. Then Miller started reading: Nina Teicholz's **The Big Fat Surprise,** Jason Fung's **The Obesity Code,** and my books. With a biochemistry background from college, Miller thought he was a reasonable judge of what was sensible and what wasn't, and LCHF/ketogenic eating made sense to him. So he tried it. Three years later he was down to 210 pounds, eating eggs, meat, and green vegetables. (He'd like to get back to 190 pounds eventually, although, he said, "that may be delusional.")

As a dentist, Miller has a unique perspective: He sees the damage to the teeth and gums that comes with eating processed carbohydrates and sugars. Many of his patients, he said, are prediabetic or have diabetes; some have gout and sleep apnea or "a whole list of metabolic disorders," and "their oral cavity is breaking down." Since Miller typically sees these patients at least twice a year, and he has plenty of time to talk to them while he works on their teeth, he does. They're a captive audience, and he believes they are a receptive one. They notice Miller's weight loss and ask him how he achieved it. It was not the result of a diet, Miller tells his patients, but rather a new way of eating that he would adhere to for the rest of his life.

In the traditional world of healthy eating, cake and ice cream **in moderation** are perfectly fine, particularly as a celebration. But when we are eating to treat a metabolic disorder, we must maintain that treatment for life. That state of remission, whether from obesity, diabetes, hypertension, or one of the many other disorders that LCHF/ketogenic eating seems to resolve, at least anecdotally, must be adhered to without limit. A little bit of cake and ice cream won't hurt you; that's not the point. You have to work hard to retain your discipline. If you indulge too soon, no matter how traditionally "healthy" the treats, no matter how whole and organic the food, how few the ingredients, you put your personal progress at risk. This, again, is one reason many of the physicians I interviewed who have clinics dedicated to weight loss and maintenance and treating diabetes with LCHF/ketogenic eating often talk in terms of breaking an addiction to carbohydrate-rich foods. There's a good reason, as Miller said to me, why alcoholics don't celebrate the successful completion of a twenty-eight-day rehab program with a champagne toast.

4. "If you do fall off the wagon, at least you know there's a wagon to get back on."

These words of advice, obviously, also come from the drug and alcohol addiction world. I heard them from Katherine Kasha, a family medicine doctor in Edmonton, Canada. Katherine is a lacto vegetarian— she avoids meat, fish, and fowl but not eggs and dairy—and has been since birth. She described herself as having always had a weight problem. "I still do," she told me. At one point she lost fifty pounds by the conventional approach and then had what she calls "an epic regain," and put on 110 pounds in four years. "I remember my husband saying, 'How is this happening with what you're eating?'"

Kasha came to LCHF/ketogenic eating through social media and then by reading Jason Fung's work. "It clicked," she said. She gave up her morning oatmeal and ate instead "a lot of eggs, a fair bit of cheese." Cottage cheese became a primary source of protein for her, in addition to high-fat dairy and tofu. She eats "lots of veggies." She makes her own yogurt and "dresses it up with chia seeds and pumpkin seeds, and unsweetened coconut." She bakes still, although now with almond and coconut flour rather than wheat flour. She sweetens with a combination of erythritol, stevia, xylitol, and monk fruit. "I'm not sure about this other stuff," she says, "but I am certain sugar is bad for me."

Kasha doesn't necessarily recommend her vegetarian eating to her patients. "I'm still going to tell you if you don't have an ethical issue, eat the damn meat," she said. "You get bonus points if it's grass-fed. It's the easiest way to get a nutrient-rich meal. There are a lot of ways, though, to do this right. There is no one way."

On LCHF, Kasha has lost significant weight, but she still finds it difficult to maintain, an ongoing struggle to avoid slipping off the wagon, particularly during holidays and family celebrations. She knows LCHF/ketogenic eating works for herself and others. "I've had patients who do incredibly well," she said. "It's phenomenal what can be done." But she wants them to understand that commitment doesn't preclude falling off the wagon on occasion. The important thing is remembering that the next step is to get back on the wagon. It's to go back to abstaining from sugars, grains, and most starches.

Andrew Samis, a critical care physician and assistant professor in the surgery department at Queen's University in Ontario, told me that he communicates the same message to patients using a cigarette-related metaphor. "One of the things you talk about when you do smoking cessation training," he said, "is what to do when you find yourself smoking again. What tends to happen in human nature is people quit smoking for six months, then slip and go back. One of the smoking cessation doctors gave me a metaphor I can now use with my patients: Every day I drive to work, I try to hit as many green lights in a row as I can. But if I hit a red light, or even five in a row, I don't turn around and go back home. I just try again from there. I still try to see how many green lights I can hit. For smoking cessation, they say as soon as you can muster the willpower to drop the cigarette and stamp it out, do it and keep trying to be a nonsmoker. What we're talking about here is a similar issue. Once people understand it's still cognition over desire, they can go back to being someone who doesn't eat these foods."

5. "It's not a religion; it's about how I feel."

This message comes from Carrie Diulus, a spine surgeon in Akron. Ohio. "I won the genetic lottery," she told me in jest. "I have celiac disease. I should be a 350-pound, insomniac, acne-laden person, but I have figured out how to work with my genetics." Extreme obesity runs in her family, Diulus said. She became a vegetarian when she was twelve years old. Her motivation was a combination of concern about animal welfare and thinking that a very-low-fat diet, as prescribed then by Nathan Pritikin, was the single best thing she could do for her own health. Her life since then was a continual series of self-experiments as she looked for a way to eat that would keep her healthy and at a healthy weight. Diulus's experience is an extreme example of the challenge.

She told me that she entered college at a normal weight, "but the freshman fifteen [pounds of weight gain] for me was more like the freshman fifty." Over the next twenty years, she yo-yoed in and out of obesity and diets. She was one hundred pounds overweight when she finished college and entered medical school. She managed to bring her weight back down eating a plant-based, calorie-restricted diet and exercising obsessively, eventually running marathons. "I was working out about twenty hours a week," she says, "and significantly limiting my calories and was able to be lean at that point, but, boy, any deviation from that, and weight came on really easily." She gained sixty pounds when pregnant with her daughter, and a "bunch of weight" during pregnancy with her son four years later. She tried a low-fat, calorie-restricted vegan diet but couldn't lose weight even while nursing. "I could have won an Olympic competition in calorie counting," she said, "but there is all this hormonal stuff going on nursing a baby. There are women for whom weight just drops off when they're nursing. Not me. This is not a thermodynamic issue, it was a truly hormonal thing."

In her late thirties, while working at the Cleveland Clinic, Diulus was diagnosed with type 1 diabetes. Her doctors, following the conventional treatment plan, advised her to take insulin and then eat carbohydrates to balance out the insulin injections. If she didn't balance the insulin precisely with the carbohydrates, though, her blood sugar would crash, and that would impact her ability to function. For a surgeon, she acknowledged, "that's a disaster."

Her solution was to stop eating grains, starches, and sugars entirely, add fat to her then pescatarian (fish and plants) diet, get into ketosis, and stay there. She then transitioned to a more standard ketogenic diet, eating meat on occasion as well. Eventually, though, she found her health markers improved significantly when she avoided meat and other animal products and ate a vegan ketogenic diet. She would still have to take insulin as a type 1 diabetic, but her diet would minimize the necessary insulin dose as well as the blood sugar swings that came with the carbohydrates. Like Katherine Kasha, she recommends her patients eat meat if they have no ethical issues doing so, but she doesn't herself. Her decision is still driven by ethical and environmental concerns as well as by the simple fact that she doesn't feel healthy when she eats meat. The same, she said, is true of dairy and eggs.

Diulus maintains her vegan ketogenic diet because it works for her. She feels healthy on it. "If I feel I need to eat meat or fish," she said, "I'll go back to them. But at this point, I feel great. I'm on less insulin than ever, maintaining normal blood sugars most of the time, and my labs are all in the range I want them. I have no reason to do anything different."

Like many of the physicians I interviewed, Diulus no longer eats breakfast. She finds that she's not hungry in the morning and functions fine without it. For lunch, she'll typically have homemade kale and Brussels sprout chips with baru and macadamia nuts or a smoothie she makes herself, consisting of kale, chard, dandelion greens, arugula, sunflower sprouts, broccoli sprouts (she grows these herself), half an avocado, juice from half a lemon, the sweetener stevia, and a tablespoon of MCT (medium-chain triglyceride) oil. She'll occasionally add protein powder made from ground sacha inchi seeds or pea protein. She also adds twenty grams of pure fiber—psyllium husk powder—that slows down the digestion of the carbohydrates. "I never get hungry in surgery anymore," she told me. "I'm cranking away in ketosis, my head is clearer than it's ever been. I'm hungry when I get home in the evening, but I'm not ravenous."

For dinners, her protein source is typically tempeh, tofu (made from either soy or hemp), black soybeans, lupini beans, or nut butters, specifically almond butter. She uses avocado oil and coconut oil for cooking and olive oil for salads and cold dishes. She makes a bread from tahini and almond flour and makes her own chocolate with coconut oil, cacao powder, and chai spices. She'll snack on nori (seaweed) sheets with olive oil and salt. Cauliflower rice is also a staple. For the first time in her life, she told me, she has to work to keep weight on. "That's shocking because I am a perimenopausal woman who was morbidly obese at one point. So this is working really, really well at the moment for me. If that changes, I'll adjust. It's not a religion."

Quite a few of the physicians and dietitians I interviewed are vegetarians who found a way to eat that works for them, that they find ethically defensible or that fits their religious precepts, but also, critically, as Diulus phrased it, allows them to feel healthy. They reached that point through a process of informed elimination and self-experimentation. What makes this interesting is that a few of the physicians I interviewed are now exclusively carnivores, not through any belief that this is the healthiest diet but because it allows **them** to feel healthy. It works for them in a way that omnivorous or vegetarian eating did not. One of those is Georgia Ede, who has worked as a psychiatrist for both Harvard University and Smith College. Abstaining from grains, starchy vegetables, and sugars can be 100 percent of the solution for some people, Ede said in our interview, but only 80 or 90 percent for others, as it was for her. "The rest of us who still struggle, we have things we have to tweak."

Like Diulus, Ede comes from a family that is predisposed to extreme obesity. Her grandmother on her mother's side weighed four hundred pounds, she told me, and all the women in her immediate family have struggled with their weight. Ede was overweight as a child and said that until about age forty she was "always on a low-calorie, low-fat, high-exercise" regime. As she got older, she found it increasingly difficult to maintain a healthy weight, even as she progressively ate fewer calories and ran more miles. In her medical residency, she no longer had the time or energy to keep up with the combination of running and semistarvation, and her weight ballooned to 190 pounds.

After her mother lost ninety pounds following Atkins's advice, Ede began to explore variations on

LCHF/ketogenic eating herself, beginning with the South Beach diet, which seemed to her the healthiest. Eventually she found she could maintain a healthy weight with LCHF/ketogenic eating as long as she avoided dairy. "I found that dairy products make me hungry and gain weight," she told me. "And for me, I could not eat as much as I wanted. I still had to be careful and I still had to exercise. But I was able to keep my weight from fluctuating if I ate the right foods. I found my groove with low-carb."

In her early forties, though, what worked for Ede stopped working. Her weight remained stable, but she developed migraines, fatigue, and concentration problems, as well as irritable bowel syndrome. She told me she gradually became incapacitated. She started keeping a food-and-symptom diary. She was working at Harvard at the time, with access, she said, to "great doctors, specialists of all kinds." But none of them asked her what she ate. So she started a series of diet experiments herself, avoiding for weeks at a time specific foods that she thought might be troublesome and recording how she felt. Eventually she got to the point where she "was left feeling fantastic, better than I ever had, even when I was a kid, no headaches, great energy, digestion was perfect, great mental stamina."

At that point Ede was eating almost exclusively meat. "It was completely the opposite of what we've been told to do," she said. "I was floored. I was scared that the diet that corrected my health was going to kill me. And as a psychiatrist, I was fascinated with the other piece of the puzzle: Why was my mood better, my concentration, mental energy, productivity? Why had my depression and anxiety gone away? It never crossed my mind that food could affect the brain like that, and yet for me, at least, it had." Her reading of the literature convinced her that her meat-based diet was safe and healthy.

In her early fifties, Ede once again made slight revisions to what she ate in response to changes in her health. "I developed some perimenopausal symptoms," she wrote to me in an email, "and some of my old familiar symptoms returned as well, including weight gain. In an effort to address these issues, I removed all remaining plant foods from my diet." Eating a plant-free diet, she said, has so far resolved all her health problems, and she lost twenty-four pounds, all the weight she had gained, plus some. "Everyone is so different," Ede said. "There are indeed basic principles that apply to everyone, but many of us have foods we're sensitive to. And we have to identify those on our own."

6. "Weight loss and weight maintenance are learned skills. You have to practice."

This is the message that Sue Wolver tries to reinforce in her patients. Wolver is the Richmond, Virginia, physician we met early in this book who changed her practice to prescribing LCHF/ketogenic eating after visiting Eric Westman's clinic at Duke and seeing his successes. This message needs little elaboration. Wolver told me that she often asks her patients if they would expect to be good at anything in life without practicing, and if they would expect to stay good at it without continuing to practice. "We have to practice at anything to get good at it," Wolver says. "And the more we practice, the better we'll get, and the easier it will be. You have to put time and effort into developing the skills necessary to do it well for life."

This is not practicing to be hungry and living with it, as the conventional thinking has always implied. It's practicing the skills necessary to avoid the foods that make us fat and sick while cooking and eating in a way that brings us pleasure. It's practicing the skills necessary to identify the foods we can eat and the foods we can't, the foods that trigger craving and the foods that trigger weight gain. It's practicing the mental skills necessary to remember, when we're craving carbohydrates, how bad we used to feel when we were eating them, how many pounds heavier we were, how much less healthy we were, and whether the gratification of a doughnut or a beer will be worth the risk of revisiting that experience.

17

The Plan

Abstaining from starches, grains, and sugars
(and replacing them with fat) takes practice, preparation,
and, ideally, the help of a good doctor.

I'd like to be able to give a detailed, very specific recipe for how to more or less rigidly abstain from starches, grains, and sugar and keep it up for life—a series of steps that guarantee success, ideally tried and true for everyone—but no such thing exists. We all have much in common, but we start from different baselines, different culinary cultures and family practices, and we have different needs. Each of us will have to pull different levers. The basics are clear—what foods not to eat—but the details will vary. Keep in mind that the goal is not to think of this as following some specific diet—i.e., "doing" LCHF or keto or paleo or some other variation—but understanding how to eat so that you can correct your weight and health by working with your physiology, not fighting it.

The advantage we have today is that LCHF/ketogenic eating is no longer a fringe endeavor. The orthodox may still think of it as a dangerous fad diet, but it's catching on because it works. It's become so common that LCHF/ketogenic foods (cauliflower rice, zucchini noodles) are now widely available in markets, in restaurants, and from online distributers. Keto-friendly processed foods are also becoming ever more widely available—shakes, candy bars, snacks that are paleo and vegan friendly as well—although I'm less sanguine about the benefits of consuming them regularly. The Internet in all its manifestations has made it effortless to obtain information, recipes, and advice (some reliable, some less so).

Thinking about LCHF/ketogenic eating—abstaining from carbohydrates and replacing those calories with fat—divides up conveniently into five ideally essential elements. For many of you, these LCHF/ketogenic eating concepts will be intuitively obvious, and the process of changing how you eat will be easy. You'll feel better. You'll have faith you're doing the right thing and can keep it up for life, a long and fulfilling one. For those who need more guidance, here are the keys:

1. Guidance. Finding a physician with whom you can work.

Abstaining from carbohydrate-rich foods and beverages and transitioning to LCHF/ketogenic eating literally changes the energy source on which your body runs, from mostly carbohydrates to mostly fat. This is not a minor transformation. You are, quite literally, changing how your body fuels itself, and it helps to do this with the guidance of an informed medical professional. With luck this process will be easy, but there's no guarantee. The odds, though, are very good that your physician is still thinking about obesity and fat accumulation (and their relationship to the chronic diseases that associate with them) along conventional lines. So it's best to find a physician who is well informed about LCHF/ketogenic eating, or who is at least open-minded and willing to do the necessary homework.

You may be on medications—for blood sugar and blood pressure, specifically—that will have to be discontinued or decreased once you have changed your diet. If this is the case, you certainly need a physician to help. He or she can do preliminary blood tests and a thorough checkup, not just monitoring LDL cholesterol but making all the assessments related to insulin resistance, metabolic syndrome, and your other health issues. If you're depressive or your hair is falling out or you have eczema or even toe fungus, it would help to have that documented in advance so you will be aware of everything that might change with this change of eating. A careful checkup can give you (and the physician) benchmarks from which to measure your progress. You can google "keto [or LCHF] physicians near me" and see what comes up, or you can go to sites hosted, for instance, by dietdoctor.com, LowCarbUSA.org, or lowcarbdoctors.blogspot.com to get started.

If you cannot find a sympathetic physician or dietitian, one possibility is to use a program like that offered by Virta Health, so that you will at least have an informed physician handy on the other end of a telephone.

2. Goals. Establishing reasonable objectives.

Setting a reasonable goal is a necessity for any successful project. If you're going to change how you eat for a lifetime, identifying your goal in advance will make it easier. What are you working for and why? Keep in mind that this is a lifetime pursuit, so your goals have to be realistic. Losing weight and getting healthy aren't perfectly correlated with happiness, for instance. They should help, but they will not be sufficient. For those who have type 2 diabetes, controlling the disease with minimal or even no medications might be a worthwhile goal, even without significant weight loss. Not having to manage a chronic disease—with all the medications and attendant costs, both human and financial—or at least making disease management as easy as possible should be very much worth a lifetime of abstaining from doughnuts, bagels, and beer. (In his 1962 memoir **Strong Medicine,** the New York cardiologist and LCHF/ketogenic eating advocate Blake Donaldson put this in his typically forthright manner: "You are out of your mind," he wrote, "when you take insulin in order to eat Danish pastry.")

You have to establish in advance a minimum amount of time you will commit wholeheartedly to this new way of eating. You have to believe that what you're giving up is worth what you're gaining back in health and perhaps weight loss. A few weeks are not enough, just as they're not enough to know whether quitting smoking or alcohol is worth the effort. I would say three months is a minimum, ideally six—long enough to get a realistic feel for what is possible.

Without clinical trials to tell us about its long-term risks and benefits, the best you can do is give it a try and see what happens. My favorite statement to this effect comes from the British Columbia physician Martin Andreae. He says that he tells his patients to embrace it for at least a month, ideally a few: "If this is bad for you, it's not going to harm you in a month or two. If you don't eat potato chips for a few months, you're not going to die from it. And if there are no disease benefits after that time, then stop if you want. But I've never had people come back and say I feel worse. I am now confident in this dietary change that everyone will benefit."

One guarantee can be given, though: If one day you determine it's not worth it and you go back to eating sugars, grains, and starches, the benefits you accrued will be lost. A lifetime of benefit will come only from a lifetime commitment.

3. Abstinence.

How best to begin will depend on what you decided were your goals and objectives. If you are going to treat the carbohydrate-rich foods you eat as an addiction, which is reasonable for many of us, then there are multiple ways to quit. In 2014, for instance, when I realized my caffeine addiction had become counterproductive, I decided it was time to end it. I could have gone cold turkey, but I didn't feel I could deal with the symptoms of caffeine withdrawal—not just the headaches but the fatigue and mental fog that would persist until my brain and body relearned how to generate the necessary mental clarity without caffeine.

So I weaned myself off slowly. A pound of coffee beans typically lasted me a week and a half.

I bought ten pounds in ten bags, and I had the local barista mix them so that the bag we labeled number one had 90 percent caffeinated coffee and 10 percent decaf; bag number two was 80–20; and so on until bag ten, which was all decaf. I went through bags one through ten in order, and it worked. I functioned smoothly throughout the transition, and by the time it was finished, I could get through a day caffeine-free without cravings. It wasn't all sunshine and roses, but it worked. I had taken a recovery/corrective process that **might** have required only a few difficult weeks and stretched it out to three and a half easier months. Which is better? I'll never know. I did what worked for me.

While the physicians and dietitians I interviewed tended to prefer the cold turkey method of carbohydrate abstinence, they were not adamant. Many said they make the decision about what to advocate to their patients or clients based largely on their assessment of what they think those patients or clients can handle, emotionally and psychologically. Can they easily embrace the necessity and concept of LCHF/ketogenic eating?

If you want to shift to this new way of eating, slowly, a step at a time, the obvious first step is to start with an essential requirement for any rational approach to weight loss, weight control, and healthy eating, regardless of the belief system: Stop eating and drinking sugar. This includes fruit juices, sports drinks, and such purported health drinks as kombucha, kefir, iced tea, and vitamin water if they are sweetened with sugar. The physicians and dietitians I interviewed will also phrase this recommendation as "stop drinking your calories," which also means no alcoholic beverages and no milk, whether dairy, almond, soy, or otherwise.

By saying this is an obvious first step, I don't mean to imply that it is necessarily an easy one. But its degree of difficulty suggests the strength of your addiction to these sugary beverages (and the caffeine or alcohol that may go with them). All the more reason to break the addiction.

All these sugary beverages are fuel sources for the body that we tend to consume between meals. They will stimulate insulin secretion, and we will burn the carbohydrates (or alcohol) in these beverages for energy during periods when we should, ideally, be burning fat mobilized from our fat cells. This step alone—**no liquid calories**—should improve insulin resistance, body composition, and perhaps energy level and mood as well. It's hard to imagine the physician and dietitian, regardless of belief system, who would not applaud and support taking this step.

One physician I interviewed who commonly recommends that his patients wean off carbohydrates step by step is William Curtis, a family physician in Corpus Christi, Texas. As Curtis told me, Corpus Christi has among the highest obesity and diabetes rates in his state, and so it's not surprising that many of his patients suffer from these disorders. Curtis was introduced to LCHF/ketogenic eating by a chiropractor friend who invited him to hear a lecture about it, which led him to attend an entire conference on the subject, and eventually he was sold. He was fascinated by the idea that multiple medical problems could be treated by nutrition alone. "I went in thinking this was quackery," he said, "but the more I listened and the more I tried it, the better the results I got. For example, I had patients who had gastric reflux and told me they'd had it forever. Once I got them to start cutting grains and sugar, they didn't have it anymore. I had diabetics who stopped eating starches and sugars, and their [hemoglobin] A1c went from fifteen [severely diabetic] down to under six [a healthy level of blood sugar control] in three months. How does that happen? There's no medicine that can do that. So I went down the rabbit hole."

Now Curtis starts his patients with what he calls the 80–20 principle: 20 percent of what we eat constitutes 80 percent of the problem. The 20 percent with his patients are sodas, sweet tea, fruit juice, and beer. "I probably gave that talk twelve times today," he said when I interviewed him in July 2017. "They say, 'What about this? What about that?' and I say, 'Just don't drink sodas, tea, fruit juice, and beer, and do that alone, and come back to see me in three weeks.' I had one

lady who lost nine pounds in three weeks just because she stopped drinking the two Dr Peppers
she was drinking every day." Once a patient sees
how much better they feel without the constant
flow of liquid carbs and sugars, he gets more as-
sertive. "You have to tell people, 'Just don't do this
anymore. It's not maybe, it's not sometimes. You
just don't do it.' And you hold them accountable.
You relate to them and tell them, 'You stop this,
you notice how you feel now. You did this! You
caused this by your choice. You fueled your body
differently and it behaved differently. Do you like
that? Yes? Then walk with me and we're going to
do some more things.'"

Another approach to the weaning process is to
remove the starches, grains, and sugars one meal
at a time. In this case, breakfast is key. A phrase
I heard more than once in my interviews is that
the typical carbohydrate-rich breakfast—cereal,
toast (with or without jam), juice, low-fat milk,
sweetened low-fat yogurt—will have you (and
your pancreas) "chasing your blood sugar" all day
long, driving moods, energy levels, and hunger
for carbohydrate-rich snacks. Switch to a breakfast of protein and mostly fat—whether eggs and
bacon or smoked salmon and avocado or some
other combination—and your insulin and blood
sugar will both remain low, allowing you to metabolize your own fat, as you'd been doing through
the night and into the morning. You should be

surprised at how satisfied you remain through the morning and well into early afternoon. Once you've altered your breakfast routine and accepted that the benefits are worth what you're giving up, you can move to lunch, dinner, and snacks. That should all be relatively easy.

For many if not most of us, though, this easy part is postponing the inevitable. It also delays the greater benefits and specifically the significant weight loss. Establishment experts often ridiculed Atkins for saying that "ketosis is better than sex," but there's a lot to be said for the energy that people experience when they are freely mobilizing fat and burning it for fuel. Until you try abstaining from sugars, starches, and grains entirely—going cold turkey—you won't know how easy it might be for you. "At the end of the day," Laura Reardon, a Halifax physician and former world-class triathlete, told me, "you want your patients to have a lifestyle that's sustainable, but you also want them to experience that paradigm shift, that 'Ahhh, okay, so this is what health feels like.' That way you're giving them both the tools and the motivation to carry on forever."

4. Contingencies. Do this right and expect the unexpected.

Any successful endeavor requires that you expect and be prepared for adverse circumstances. In this case, you want to preempt, if possible, or at least minimize the adverse symptoms of carbohydrate withdrawal that might derail your progress or seem like a reason to quit. When physicians and researchers talk about a "well-formulated" LCHF/ketogenic diet, a concept popularized by Steve Phinney, Jeff Volek, and their company Virta Health, they mean one that will minimize side effects while maximizing benefits. If you're going to switch to LCHF/ketogenic eating, you want to make sure you do it right.

The transition from burning mostly carbohydrates to burning fat has physiological effects beyond mobilizing fat from your fat cells and stimulating ketone production. Historically the orthodox authorities have wielded any adverse side effects as reasons to avoid LCHF/ketogenic eating, but they're mostly symptoms of the withdrawal process. These symptoms are no more a reason to go back to eating sugars, starches, and grains than delirium tremens would be a reason for an alcoholic to go back to drinking. Only a minority appear to suffer from them, but preparing for the eventuality and understanding the mechanisms will help you weather the storm if you do. As a physician would want you to know what to expect from any drug they might be prescribing, and what to do to minimize side effects (e.g., take the pill on a full stomach), the same is true with embracing LCHF/ketogenic eating.

The most common side effect is what used to be known as the "Atkins flu," now typically known as the "keto flu." As I said earlier, when you lower insulin, your kidneys will excrete sodium (salt) in urine rather than retain it. This goes along with several pounds of water that's no longer bound up with glucose in its storage form, glycogen. This "water weight" is lost at the beginning of any diet, whether calorie restricted or carbohydrate restricted, but it's more extreme in LCHF/ketogenic eating; the absence of carbohydrates means the glycogen stores are more quickly depleted. The combination of water loss and sodium loss appears to be a major cause of most of these flu-like symptoms, perhaps all of them, including headache, fatigue, nausea, lightheadedness, and constipation.

At its worst, the keto flu can be debilitating. Barbara Buttin, a gynecological oncologist in suburban Chicago, told me that the first time she tried LCHF/ketogenic eating, she quickly quit because of the keto flu. "I couldn't sustain it," she said, "because I couldn't function at surgery. Then I tried again a few months later, and I made it through. The second time it was like a couple of days without coffee." Other physicians said some of their patients feel "like crap" for a few days to a few weeks while their bodies adapt to mobilizing fat and using it for fuel.

There are no hard numbers on the likelihood that you'll feel these withdrawal symptoms. Some physicians I interviewed said these symptoms were common among their patients; some said they were uncommon. "I can't quite put my finger on the rhyme or reason behind why some feel it worse than others," Kelly Clark, a nurse practitioner who owns and operates two medical weight-loss and wellness clinics in southeastern Wisconsin, told me in an email, "Personally, I had a KILLER 3 day headache, had dreams of eating the tops off muffins (I don't even like muffins) and at one point I nearly cut through three lanes of traffic to turn into a grocery store that carried my favorite chocolate chip scones! Just crazy!"

The lesson from anecdotal and clinical experience is that you can avoid the keto flu or reduce the symptoms by making sure you're eating sufficient fat and, more specifically, by replacing the sodium and water you're losing. Hence, "eat salt, drink water" is now advice that goes along with abstaining from carbohydrates, even if "eat salt" is another way LCHF/ketogenic eating diverges from the conventional notion of a healthy diet. We've been told to avoid salt for the past fifty years because the nutritional authorities believe it's our high-salt diets that raise blood pressure and cause hypertension. This is another hypothesis that repeated experiments failed to confirm but was accepted as true nonetheless.

The likely explanation for why hypertension is associated with obesity and diabetes and metabolic syndrome (i.e., insulin resistance)—why high blood pressure is one of the diagnostic criteria of metabolic syndrome—is that insulin and insulin resistance influence all these disease states. Reverse the insulin resistance and lower circulating insulin levels with LCHF/ketogenic eating, and blood pressure will drop, independent of salt consumption. Add salt and water back to avoid the symptoms of the keto flu, and blood pressure should still remain low and in a healthy range. In those who have hypertension, blood pressure clearly drops—in clinical trials and clinical observation—despite the liberal use of salt in the diet.

Phinney and Volek, who have the most research experience in this field plus the clinical experience garnered from Virta Health, recommend taking four to five grams a day of sodium, which is about two teaspoons of salt, or about twice what the average American consumes (typically in the processed, carb-rich, foodlike products you won't be eating anymore). You can accomplish this by liberally salting your food to taste when eating and cooking. They also recommend magnesium supplementation—300 to 500 mg a day initially—to help with muscle cramps, which are common with LCHF/ketogenic eating and a sign of magnesium depletion.

The loss of sodium is the primary reason physicians and dietitians who prescribe LCHF/ketogenic eating suggest you drink a cup or two of broth or bouillon every day—made from the bones of meat or poultry. (It's also a reason a 2015 **New York Times** article called bone broth "a trend beverage, ranking with green juice and coconut water as the next magic potion in the quest for perfect health.") A store-bought bouillon cube dissolved in hot water will do the trick. For those who are averse to drinking either, pickle juice is another sodium-and-electrolyte-rich solution.

Two other possible withdrawal symptoms are also related to sodium depletion: postural hypotension and abnormal heart rhythm. Postural hypotension means that when you switch to LCHF/ketogenic eating, your blood pressure can drop so low, it doesn't adjust properly when you go from lying or sitting to standing. Dizziness, even passing out can follow. Adding salt to your diet, drinking broth or bouillon, and, ideally, taking magnesium and potassium supplements should resolve both postural hypotension and any abnormal heart rhythm problems. (You can now buy keto electrolyte supplements that combine all these minerals together in capsule form.) Still, it's essential with both these conditions that you see your physician or cardiologist and make sure nothing more serious is going on.

Most physicians also recommend a multivitamin with LCHF/ketogenic eating to more or less cover any bases. Meat and eggs, though, are rich sources of essential vitamins and minerals, as are, of course, green leafy vegetables, which you're likely to eat copiously at least once a day. (The fact that there's now a "carnivore" or "zero carb" movement populated by Georgia Ede and others who seem to be healthy eating nothing but animal products strongly suggests that even green leafy vegetables may not be a necessary component of a healthy diet. These people and their zero-plant diets have incited considerable controversy, but their experience cannot be ignored.)

Another withdrawal symptom that appears to be relatively rare is the exacerbation of preexisting gout. The excruciatingly painful symptoms of gout are caused by an excess of uric acid, and we store uric acid in our fat cells. The uric acid, too, is mobilized when insulin is low, and the excretion of uric acid in our urine uses the same kidney transport system as ketones. ("By competing with uric acid for renal tubular excretion, elevated blood ketones can promote hyperuricemia" is how this was described technically in the 1973 AHA critique of Atkins.) Raise ketone levels, and uric acid can accumulate in the circulation and lead to an outbreak of gout. Your physician can handle it as he or she would any gout outbreak. Eventually circulating uric acid should return to a healthy level, and the gout, too, should be temporary.

Experiencing any of these effects of carbohydrate withdrawal can feel like a reason to go back to eating starches and grains. But the symptoms should not last. The worst case is typically "a couple of weeks of feeling like crap," as Patrick Rohal, a physician in Lancaster, Pennsylvania, told me. The solution is drink water and broth or bouillon, add salt, take magnesium supplements (if necessary), and have patience. If the withdrawal symptoms don't abate, work with your physician to find out why.

The one side effect of LCHF/ketogenic eating that may be lasting is the one that is likely to make physicians most anxious. This is the effect on LDL cholesterol, the "bad" cholesterol, as it's known in the conventional thinking. As discussed earlier, the common wisdom on a healthy diet is driven disproportionately by thinking about this single number—LDL cholesterol—and the unwarranted belief that it is a strong predictor of heart disease risk. Physicians learn in medical school that if LDL cholesterol is high, patients should be prescribed cholesterol-lowering drugs called statins. And, of course, if their patients are eating diets with even moderate amounts of fat, they should stop doing so.

The role of LDL cholesterol itself is still controversial. (It can be controversial even to say that it's controversial.) The relevant point is that something about eating fat-rich diets devoid of refined grains, sugars, and starches can increase LDL cholesterol and also drive up the number of LDL particles, which is a far better predictor of heart disease risk, as I've said. The data do not exist to say what proportion of people will experience elevated LDL (cholesterol or particle number) when they abstain from carbohydrates, but it is not a rare experience. It may be caused by the saturated fat content, although people eating paleo diets, which tend not to be saturated-fat-rich (dairy products and butter are not strictly paleo), have also been known to experience elevated levels of LDL. In the LCHF/ketogenic eating world, these individuals are now known as "hyperresponders," and the only reliable way to know if you are one is to abstain from eating carbohydrate-rich foods and find out.

The more important questions are about whether this is harmful and, if so, how harmful: (1) whether elevated LDL (cholesterol or particle number) means you are indeed at high risk of having a heart attack even if you're abstaining from carbohydrates, (2) whether that risk is significant, and (3) whether the benefits of correcting obesity, diabetes, and all the metabolic disorders associated with them and with insulin resistance (hence, metabolic syndrome) are offset by this increase in LDL cholesterol.

It's not uncommon for physicians, confronted with a patient who is following LCHF/ketogenic eating and whose LDL cholesterol has gone up, to talk that patient back to carbohydrates—to

suggest, in effect, that they go back to eating potatoes and toast, as the Irish physician Daniel
Murtagh put it—regardless of how much weight
they might have lost and how significantly their
blood pressure and blood sugar control might have
improved. To several generations of physicians,
keeping LDL cholesterol low is the be-all and end-
all of heart health. Now, however, as physicians
become more informed on the benefits of LCHF/
ketogenic eating, or at least more open-minded,
this kind of knee-jerk conservative response is
less common.

At the moment, this question of trade-offs—
lose weight, control blood sugar, and lower blood
pressure but elevate LDL in the process—is mired
in controversy, and there are no long-term trials
capable of resolving it. Informed physicians and
researchers I've interviewed (and by "informed," I
mean those who have made an effort to understand
both sides of the science) would argue that even
in the worst-case scenario, in which LDL matters
tremendously and it goes up, you can still bring it
down either by reducing the saturated fat you're
consuming and replacing it with monounsaturated
fat (exchanging butter for olive oil, for instance) or
by using a relatively benign cholesterol-lowering
drug, or both. Either way, you can still keep all
the other health benefits that come with abstaining from carbohydrate-rich foods.

If your physicians lean toward conventional

thinking, they will advocate that you take a statin or other cholesterol-lowering drug if your LDL is elevated. If they don't, it might be because they assume, as I do, that possible side effects from the drug, one that you are expected to take daily for the rest of your life, are not worth whatever small benefit in longevity and health it might provide. (Sarah Hallberg, who ran the trial for Virta Health at Indiana University on individuals with type 2 diabetes, says in their experience "hyperresponders" who are in nutritional ketosis can lower their cholesterol with low-dose generic statins that cost dollars a month and seem otherwise benign.)

The last time I had my own "blood lipids" assessed, which was several years ago, my LDL cholesterol was elevated, as was the number of LDL particles in my bloodstream. They hadn't been a few years before that. But I have no other risk factors for heart disease. I chose to live with the high LDL and avoid taking a drug for the rest of my life that would have no short-term benefits on how I feel. It's an informed gamble. I don't like being dependent on a drug, particularly one that is supposedly preventing disease in the future, not addressing symptoms in the present. I'm willing to take the risk. I have physician friends who are highly informed on LCHF/ketogenic eating, whose opinions I respect, and who have high LDL-P (the particle numbers), and they do take low-dose statins. It's a judgment call you have to make that will depend on your sense of the evidence, and your trust in your physician and whatever she or he recommends.

5. Adherence. Sustaining what you started in a world that makes it as difficult as possible.

When I interviewed Jeff Stanley, a physician in Portland, Oregon, who now works with Virta Health, this is how he described the two primary challenges of prescribing LCHF/ketogenic eating to his patients: "The biggest challenge in getting people to succeed with this way of eating is getting them to try it. Once they try it, they'll experience the benefits. But then the second-biggest challenge is getting them to stay with it." That's not because the benefits diminish or disappear— the patients are still losing weight without hunger and they feel healthy—but because, he said, "life circumstances get in the way." Stanley says he has patients who abstain from carbohydrates, lose fifteen pounds in a month, "feel awesome," and then go on vacation, fall off, and have trouble getting back on. "It's getting people to adopt it as a lifestyle that's important. People have to realize how much better they feel when they follow it, how much healthier they are, even if it means staying away from bread and cupcakes."

One challenge of abstaining from carbohydrates

that doesn't exist quite so intensely in quitting cigarettes, for example, is that the world conspires to make carbohydrate abstention as difficult as possible. You can't leave your house or turn on the television (or get on the Internet) without being tempted to step off the wagon. Every holiday, every dinner out, every occasion with friends, every office get-together or even coffee break, is an experience in having to say no to foods and treats you no longer eat but your friends, family, and coworkers do. This experience will be driven by Pavlovian and cephalic phase responses creating an urge to eat these foods that will then manifest as a rationale to eat them, "just this one time."

When you quit smoking, your friends are very likely to help and encourage you along the way. If they smoke, they will try not to smoke around you. They won't say no if you ask them to take their cigarette outside (or at least your real friends won't). Many governments now help by enforcing rules against smoking in public spaces. Even if the stated rationale is to protect us from secondhand smoke, one consequence is to make it easier for smokers to quit. If you're abstaining from carbohydrates, though, you can't expect your friends not to order pasta if you eat at an Italian restaurant—they're not abstaining—or say no to a dessert because you don't eat them, or not to have a birthday cake at **their** birthday party. Few of us like being the person at the party who's saying no to the treats or the cake, but it's a skill we have to learn.

When I asked Garry Kim, the physician who runs a chain of weight-control clinics in the Los Angeles area, how he handles this challenge with his patients, he said that, having no control over it, he tries to demonize the food environment. "I try to impart an us-versus-them mentality," he told me. "People are conspiring to make us all fat, and we have to fight against it. We have to draw a line and not let them win."

Losing this battle can be all too easy. One of the newspaper editors with whom I occasionally work described this phenomenon to me from her personal experience after she began LCHF/ketogenic eating. "One little slip, and I'm back on carbs," she said. "It's like I eat nothing, then have a grain of rice, and before I know it, I'm snarfing a doughnut."

For many of us, the sensation of being on the edge of a slippery slope is ever present. This is why I personally find it easier to avoid sugar entirely than to try to eat it in moderation. Eating a few bites of a tasty dessert doesn't satisfy me (as it apparently does my wife); rather, it creates a craving to eat it all and then some. Allowing myself to eat grains and starches in moderation also makes me hunger for more. Eating fat-rich foods "helps extinguish binge behavior," as the Harvard pediatrician and nutritionist David Ludwig (author of **Always Hungry**) phrases it, "as opposed to high-carb foods which exacerbate it." The insulin dynamics discussed earlier help explain this phenomenon.

Whatever the mechanism, if the goal is to avoid the kind of slip that leads from "just this once" or "just one bite" thinking to binge behavior and returning to a carb-rich, sugar-rich diet, then the same techniques that have been pioneered in the field of drug addiction for avoiding relapses should work in this scenario as well. These basic principles have evolved over decades, and addiction specialists believe they can work for anyone who's gotten "clean and sober" and wants to stay that way.

Many of the principles are common sense, the kind of advice we would give our children if we were trying to keep them away from trouble. If we're trying to avoid a source of temptation, then we do our best to ensure it's out of sight so it has a reasonable chance of being out of mind as well. "Alcoholics who care about staying sober won't get a job in a bar or even walk down the alcohol aisle in a grocery store," the University of California, Berkeley addiction specialist Laura Schmidt told me when I interviewed her for a story about breaking an addiction to sugar. It's harder to avoid the carbohydrates that trigger our cravings because they're more ubiquitous in our environments, but we have to work to make that happen nonetheless.

It will help, for starters, if you clean out your immediate environment—your kitchen and your cupboards, maybe even your desk drawers— so that it's free of the kind of carbohydrate-rich foods that tempt you. In 2013 David Weed, a psychologist in Fall River, Massachusetts, started a community health program that was awarded a Culture of Health Prize by the Robert Wood Johnson Foundation. His program included a ten-week course on LCHF/ketogenic eating as part of an annual fitness challenge that attracted over a thousand people each year. More than a hundred people took the course over the five years he offered it. He said the ones who succeeded in his course— the ones who "do it best"—were those who literally did a carbohydrate cleanout at home first. They then filled the refrigerator and freezer with the kinds of ketogenic-friendly foods they could cook and eat anytime. He told the participants that they had to respect the power of their environment: "If you bring any food into the house, you're going to eat it, no matter whether you should or not. Don't think that you have the willpower to not consume carbohydrates that you bring home. The decision point has to be at the grocery store." "You buy it, you wear it," he would say.

This means you have to plan ahead for experiences and environments that are likely to weaken your resolve. Among the habits you have to form and reinforce is that of thinking ahead about what you can and cannot eat at office parties, in airports and on flights, on vacations, and at holiday meals. If you expect no LCHF/ketogenic-friendly foods will be available, then plan to bring your own. If you were a vegan or a vegetarian, this is how you would approach such situations without a second thought. **What can I eat?** It's a natural question to ask yourself if what you eat is not necessarily what everyone else is likely to be eating. As Carrie Diulus told me, she won't get on a plane without a bag of macadamia nuts for her snack. If everyone else is eating and no LCHF/ketogenic options are available, it helps to have one in your bag. This kind of thought and effort would go into any attempt to eat healthy. It simply requires a tighter, different focus.

A greater implication of this power-of-the-environment phenomenon is that you will more likely succeed if your family eats the same way you do. If you believe it's the healthiest way to eat, it will help if everyone at home agrees with you. A smoker is less likely to quit in a house full of smokers than in one in which all are quitting together or no others smoke. The same is true with LCHF/ketogenic eating. "People who are most successful in making the change," Weed said, "are people who have a spouse who has also bought in. People with the biggest challenge are the ones who go home to a household full of carboholics, including kids and a spouse who insist on consuming high-carb meals. For many, it's just too difficult to manage, and they give up, not because low-carb isn't working but it's just too hard to stay with it in that kind of environment."

Since your environment includes your social network—your friends and coworkers—it will help if they at least understand what you're doing and support it. Changing your social networks may be necessary as well. Beyond convincing your family and friends to be invested in your health, just as they would be if you were trying to quit smoking or alcohol, you may need to find an LCHF/ketogenic eating group to join (online, if necessary) so that you have a community that is supporting what you're doing, can help with questions or give advice, and can help you get back on the wagon when you fall off. It's the same reason alcoholics go to AA meetings and people with other psychological and addiction issues go to group therapy sessions. "I have never had anybody who tried this who didn't get good results," Weed told me, "not one. I have lots of people who get good results and then trip up. I always ask 'Why did you stop?' I'll get a whole range of vague answers, but it largely comes down to the fact that people get little support for it. They do well if they're part of a group doing it. It's really an important part of the practice: people doing it in a group setting, they learn enough about it, and, more important, they learn from their peers who are doing it, too."

This is all part of the process of going down the rabbit hole. Not only can you follow discussions on Twitter and Instagram on LCHF/ketogenic eating, but you can follow websites like Dietdoctor.com and Diabetes.co.uk and join Facebook groups as well. Even those who eat no meat and animal products can join multiple Facebook groups of vegans who follow LCHF/ketogenic eating, one of which has over fifty thousand members as I write this, and thereby get support, advice, recipes, and help when needed.

6. Experimentation. Knowing which levers to pull when LCHF/ketogenic eating does not sufficiently correct your health.

LCHF/ketogenic eating will work "as if by magic" for some of us and not nearly as well for others. Some lose all their excess fat, and some not nearly as much as they'd like. Some resolve all their health problems, and some don't. Some get lean and healthy for a while, even for years, and then good health and a healthy weight become more elusive.

Here's where individual variation is the critical factor. Some of this is clearly due to the action of hormones other than insulin on fat accumulation (though insulin remains the dominant link to what we eat). This is why men seem to have an easier time losing excess fat than women, and

younger men and women will have an easier time
than older ones. This observation was first made
by the British physician Robert Kemp in a series of
articles reporting on his clinical experience recommending LCHF/ketogenic eating to over fourteen
hundred overweight patients between 1956 and 1972.
Many of the physicians I interviewed, although
not all, agreed with it. Both testosterone and estrogen suppress fat formation, testosterone above
the waist, estrogen below it. As we get older and
secrete less of these hormones, this inhibition is
weakened, and our fat cells respond by getting fatter. Some of us will only be able to lose some of
our excess fat; others will fare much better. Insulin
is the dominant hormone in fat accumulation, the
one we can manipulate most obviously by changing our diet, but that may not be sufficient.

In the course of my interviews, when I asked
when and why LCHF/ketogenic eating fails, why
some people do not lose weight and some may
even gain weight, many of those I spoke with had
a story—if not about their patients, then about
themselves. Carrie Diulus, for instance, said that
she personally gains weight when she consumes
too much butter. She doesn't have an "off switch"
with butter, she said, so she's learned to avoid it.
She also limits her access to ketogenic "treats" (as
do I), such as low-carb cakes, cookies, and other
desserts, and eats them only occasionally. It is just
too easy, she says, to eat these foods to excess. A

psychologist I interviewed who has been running a series of clinical trials comparing LCHF/ketogenic eating to more traditionally "healthy" diets (and who requested anonymity) told me that she'll find herself craving carbohydrates if she eats as few as four strawberries a day. She therefore avoids strawberries, though she still eats other berries in season. I find that once I start eating nuts, I crave them, and so my best strategy and my healthiest weight come when I avoid them. Because I'm confident I have the skills and habits necessary to remain stable at a healthy weight for the long term, and because I'm very fond of nuts, I allow myself to fall on and off the nut wagon.

These kinds of anecdotes speak to what these physicians describe as the need to "find your triggers," not just the environmental triggers that endanger falling off the wagon but the foods that inspire craving, that others might be able to eat with impunity but that you apparently can't. While the basics of LCHF/ketogenic eating are obvious—abstain from grains, starchy vegetables, and sugars, replace those calories with fat—individual variation is where the plateaus come in, and tweaking or massaging or fine-tuning what we eat (and don't eat) is necessary.

Without the help of an experienced physician or dietitian, we have to examine our own diets and experiment with fixes to see what the problem might be and what works (or doesn't). The experiments themselves should be at least several weeks long to give them time to work. The problems fall into three main categories, or three levers to consider pulling when LCHF/ketogenic eating is no longer working or not working well enough.

First, the obvious one, is determining if you're still eating too many carbohydrates. With patients who are weight stable yet swear they are religiously avoiding carbohydrates, physicians will have them keep a detailed food diary for three days—there are now apps that will make this relatively easy—so they can then see if that's true. This would also be a good time, physicians told me, to check if the patient is in measurable ketosis, not so much because ketosis is necessary to achieve health and a healthy weight but because it will be a reliable sign of carbohydrate restriction. If the patient is in ketosis, the physician can have faith that he or she is indeed abstaining from carbohydrates and then move on, if necessary, to other possible explanations.

It's important to examine how carbohydrates might be sneaking into your diet unnoticed: the apple a day that you still consider a healthy snack, the cornstarch used to thicken gravies, the accumulation of carbohydrates from nuts and nut butters. (Sue Wolver told me about a patient with diabetes whose "beautifully controlled" blood sugar levels with LCHF/ketogenic eating suddenly went awry. This patient had an upset stomach at one point and started popping TUMS almost hourly. TUMS is an antacid for acid reflux, and each tablet contains about 1.5 grams—6 calories' worth—of carbohydrates. This was enough to significantly worsen this woman's blood sugar. Once Wolver discovered what was happening and her patient discontinued the TUMS, her blood sugar levels returned to normal.) Another common problem, as Ken Berry puts it, is confusing the concept of "less bad" with "good." "I have patients," he told me, "who eat sweet potatoes because they've heard they're not as bad as regular potatoes. Or they'll tell me they're not eating bread, but they're eating flour tortillas instead or whole wheat bread instead of white bread. All this might be less bad, but it's not good enough."

To be sure, a lot of this ultimately is common sense. (Or at least it's common sense in a world in which obesity is a hormonal/regulatory disorder and the link between fat accumulation and what we eat runs through insulin and carbohydrates.) If you're buying low-carbohydrate products in the supermarket, for instance, and you're not losing weight, these packaged foods might be the problem. Brian Sabowitz, an obesity medicine specialist in Spokane, described this to me as "you think you're on a low-carb diet and you're not." Sabowitz said his favorite example was tuna salad purchased as a prepared food in the supermarket. "If you don't happen to look at the label, you think you're getting tuna and mayonnaise and maybe little celery chunks. When you look at the label you see another ingredient is high-fructose corn syrup. You're getting a bunch of sugar in your diet, and you think low-carb didn't work because you did it and it failed."

Few of the physicians I interviewed believed the concept of "net carbs" was a useful one. Net carbohydrates is a measure of only the carbohydrates that are digested and absorbed into the circulation; it doesn't include the carbohydrates that we don't digest and metabolize (the fiber). Net carbs can be useful to ensure you're staying under some predetermined maximum of carb consumption every day—say, the 50 grams often set as a limit for ketogenic diets. But if your weight loss has stalled while you still have excessive body fat, then it might be a good idea to trust your body rather than the manufacturer's determination of net carbs. The goal is more or less rigid abstinence, and your body will tell you if you're erring on the side of being too liberal.

Second, too much fat in the diet has also ironically become a possible reason for weight loss to stall. Now that LCHF/ketogenic diets have moved toward the mainstream, they've been accompanied by new ways to infuse the body with fat that might be physiologically unnatural. Until recently, humans rarely if ever had the opportunity to drink fat without at least some protein or carbohydrates to go with it. Now we do. Bulletproof coffee, for instance, popularized by the Silicon Valley entrepreneur Dave Asprey, is a mixture of coffee, butter (or ghee), and MCT oil, typically derived from coconut oil. Because the MCT (medium-chain triglycerides) are metabolized primarily in the liver, they can increase the synthesis of ketones even with some carbohydrates in the diet. This is why bulletproof coffee may provide an energy boost beyond that of the caffeine in the coffee alone, but it also floods the body with fat, or at least trickles it in like an intravenous infusion over the course of hours.

Some of us may be fine with that. Some of us may burn the fat we're eating (or in this case drinking) and still have excess fat that remains stored in our fat tissue at the end of the day. "I can eat a pound of coconut oil right now," Ted Naiman, a Seattle physician who has been advocating LCHF/ketogenic eating for almost twenty years, told me, "and I will be in the deepest ketosis you've ever seen, and I won't lose any weight. I will be burning fat, but it's the fat I ate, not the fat I stored." Whether your body burns or stores these fat infusions can also change with time. What we can tolerate during weight loss may be different from what we can tolerate once we're weight stable. The idea that we can eat as much fat as we want without storing some as excess may be true for some of us but not for all.

Third, you may be consuming too much protein.

It's a common tendency, as we discussed earlier, to try to compromise between health paradigms by eating a low-carbohydrate diet that is also a low-fat diet. (This is what many physicians used to prescribe for obesity, pre-1970s, because they thought the body needs protein but can live without the calories from both carbohydrates and added fats.) The result, notwithstanding the unsatisfying meals, is calorie-restriction or a high proportion of protein in what you're eating. The amino acids in the protein can elevate insulin in response, which might be enough to stimulate fat accumulation and hunger, including cravings and binge behavior. The solution is to add more fat: butter or olive oil on the vegetables; chicken thighs with the skin on instead of chicken breasts without; fatty cuts of meat and fatty fish instead of lean.

The use of artificial sweeteners may be another reason LCHF/ketogenic eating can work poorly. Most of the physicians and dietitians I interviewed think of these sweeteners, as I do, as a useful crutch while transitioning to LCHF/ketogenic eating and breaking a sugar addiction—the "methadone of sugar," as Sue Wolver calls them. Artificial sweeteners from "natural" sources—stevia, for instance, which comes from a shrub in Central America, or monk fruit—may be more benign than the sweeteners that were invented or discovered in a chemistry lab in modern times. But that's an assumption for which we have no meaningful experimental evidence. Saccharin, which was first discovered in coal-tar derivatives, has been used as a sweetener since the 1890s. It's three hundred to five hundred times as sweet as sugar, which means obtaining the equivalent sweetness requires only 1/300th to 1/500th the dose. It also passes through the body without being metabolized, a good thing. The evidence that these artificial sweeteners are harmful in and of themselves is not compelling (to me). Some evidence exists, though, that they may fool our bodies into thinking we're consuming sugars and respond in a way that might interfere with fat metabolism and the use of our stored fat for fuel. It might do so just by making us hungrier and even hungrier for carbohydrates.

When weight loss stalls and you still have significant excess fat to lose, common sense is required. Ask yourself what you're eating or drinking that might be interfering with fat metabolism. If you're using an artificial sweetener, that's an obvious suspect: It makes sense to see what happens when you avoid it. Give it up for a few weeks. The harder that may seem to you, the more likely it is that giving it up is a good idea. If it makes a difference in how your body responds to LCHF/ketogenic eating, if you start getting leaner again, then you know artificial sweeteners trigger a response in **your body** that is problematic. Your sweetener of choice may be fine for some but not for you. You can also add it back into your diet to see if weight loss stalls again. If it does, then it's telling you that your body can't tolerate these sweeteners.

Common sense also dictates that you break yourself of your craving or need for sweets. Ideally when you shift to LCHF/ketogenic eating, you'll find pleasure in food from the savory elements—the salt and the fat. Have patience, though, as these changes in taste and preference can take time.

Alcohol presents a similar issue, in which common sense is a good mediator. If you're maintaining significant excess fat on LCHF/ketogenic eating, your alcohol intake might be the problem. Alcohol can be thought of as a fourth macronutrient. Its caloric density (7 calories per ounce) is between that of carbohydrates and protein (4 grams) and fat (9 grams). Drinking cocktails with sugary sodas (tonic water) or sugar-rich alcohols (e.g., brandy) will likely be fattening. The calories in beer are from carbohydrates (maltose) as well as alcohol. Even the carbohydrate content in low-carb beers might be enough to put you over the insulin threshold. Some of us will be able to tolerate it, some of us won't. Red wine is better than white wine, because it has fewer calories and less sugar, but multiple glasses of red wine daily (or even weekly) might derail weight loss. This is clear from anecdotal and clinical experience.

Alcohol is metabolized in liver cells much like the fructose from sugar, and it can cause similar problems—particularly fatty liver. The liver will burn alcohol and generate energy in doing so, and the heart muscle and kidney will burn the metabolic by-product of that process (acetate). If they're doing that, though, they're not using fat for fuel, and it may be accumulating. If you don't find out how your metabolism runs without the alcohol, you'll never know if it's worth the trade-off.

There's a lot to be said for living well, however we define it. But if you're drinking alcohol of any kind while embracing LCHF/ketogenic eating, and you're still maintaining excessive fat and insulin resistance, experimenting for a month or two without the alcohol seems worthwhile. (If the prospect of going a month or two without alcohol seems intolerable, as with the artificial sweeteners, practicing abstinence may be more important than you think.)

You might also try exercise, but not for the purpose of burning calories. As with everything in this world that involves fat and fuel metabolism, we have to change our perspective. From this viewpoint, physical activity is the kind of thing you want to do when you're metabolically flexible, insulin sensitive, and burning your own fat. It's not an effective way to force your body to reduce your fat stores. Burning off calories through exercise is likely to make you hungry, as we discussed; it's unlikely to make you significantly leaner.

One possibility, though, is that building muscle can help, which implies resistance training (weights) rather than burning calories through cardio. A few clinical trials have suggested that resistance training augments weight loss with LCHF/ketogenic eating. A bout of resistance training (or cardio) will deplete your glycogen stores and make you more insulin sensitive while your cells try to replace the lost glycogen. If you're doing LCHF/ketogenic eating, this increase in insulin sensitivity may be meaningful. The Seattle physician Ted Naiman says he has seen some of his patients—"older women, who are very sedentary and massively stalled out"—return to losing excess fat by going to the gym and doing resistance training. It's worth a try. Otherwise, exercise simply because it makes you feel good, if it does. That's reason enough.

One of the experiments I tried, beginning in August 2017, was intermittent fasting or time-restricted eating (TRE). The simplest way to put it, in my case, is that I stopped eating breakfast. All my meals, including snacks, now fell between lunch, around one p.m., and dinner, which typically ended by eight p.m.

The technical definitions of intermittent fasting and time-restricted eating overlap, and this can get confusing. Both terms can refer to eating only two meals a day, as I did, and no snacking after the second one: either breakfast and lunch, and then skipping postlunch snacks and dinner, or eating lunch and dinner only, and then avoiding after-dinner snacks and skipping breakfast. Hence the term **time-restricted eating** refers to the window of time during the day in which you **are** eating—say, seven hours from lunch to the end of dinner, in my case. **Intermittent fasting** refers to the time you're **not** eating: the seventeen hours, in my case, between dinner and lunch the next day.

With either term, you're extending the duration of time in which you have to rely on your fat stores for fuel. You're prolonging the amount of time you're under the insulin threshold and fat is being mobilized and oxidized. People who embrace it (under either name), as I did, say they find it easy to skip one meal a day if they're already doing LCHF/ketogenic eating, although it may take a few days to get used to it. In other words, they feel no additional hunger by not eating breakfast or dinner.

I was initially skeptical about intermittent fasting/ TRE, assuming that it was most likely a passing fad. A few years from now, I thought, we'll all be saying, "Remember back around 2018 when everybody was fasting, skipping meals, going days without eating?" Then I had three days of travel, all morning flights, which presented an easy opportunity to try it. All I had to do was say no to airplane food, which is never that difficult. By the time I returned home, not eating breakfast was surprisingly easy. Over the next few months I lost a dozen pounds that I didn't think I needed to lose, and I did so without hunger. I kept it up and still do because I feel better when I don't eat breakfast. I have more energy and mental clarity. I'm no longer hungry in the morning, and having my first meal of the day early in the afternoon feels normal to me now. I don't think of what I'm doing as particularly faddish. I'm just not a breakfast person anymore, and that's how I talk about it.

Nutrition researchers are now doing clinical trials testing the benefits of intermittent fasting/TRE, typically comparing it with other means of reducing calories. The researchers, in other words, assume fasting works because we eat less during those periods, as we do, and that's why we lose weight. As noted, it also prolongs the period that our fat cells are below the insulin threshold, experiencing "the negative stimulus of insulin deficiency" and therefore mobilizing fat. Either way, it's a reasonable assumption that intermittent fasting/TRE has become common because for many people it works, as it did for me: It makes them leaner and healthier and does so without hunger. It's not a religion, as Carrie Diulus would say, it's about how we feel. You don't need a clinical trial to tell you if intermittent fasting/TRE works for you. You can try it and find out.

The term **intermittent fasting** can also be used to refer to the 5:2 diet plan, popularized by the British physician-turned-television-journalist Michael Mosly, in which for two days a week you restrict your calories to under eight hundred a day (and your carbohydrates to under four hundred). It can also imply fasting regularly for days or a week or more, as popularized by the Toronto physician and kidney specialist Jason Fung. (Many of the Canadian physicians I interviewed for this book credited Fung's 2016 book, **The Obesity Code,** with introducing them to LCHF/ketogenic eating.)

When I interviewed Fung, he told me that many of his patients had obesity and/or diabetes and this was the reason for their kidney problems. Around 2012 he started recommending LCHF/ketogenic eating but with little success in his practice. A significant proportion of his patients were immigrants from the Philippines or Southeast Asia, and he had trouble communicating the idea to them, he said, let alone convincing them, that they should no longer eat rice or noodles, the staples of their diet. He started thinking about other ways they might lower their insulin levels without pharmaceutical therapy, and he hit on fasting.

"What's wrong with this idea of intermittent fasting or even extended fasting for seven days?" Fung said. "I started looking into it, and there's really nothing wrong. People have been doing it for thousands of years, and it has the same ultimate goal of LCHF/ketogenic diets, which is that it drives insulin low for extended periods of time. Everything is going to be maximally minimized. I read all the literature, there's nothing there that tells me people can't do it. I'm not talking about thin people going without food for forty days. I'm talking about three-hundred-pound people going twenty-four hours without food."

Fung had an easier time, he said, convincing his patients to fast regularly—anywhere from twenty-four hours, dinner to dinner, two or three times a week, as he now does himself, to a week or more for his heaviest patients. He still tries to get his patients on a **relatively** low-carbohydrate, high-fat diet, but then he adds fasting as well. He told me stories of patients who had been on 150 units of insulin a day—high doses—with severe type 2 diabetes getting off their insulin within two months. As for his track record, he said he can convince about half his patients to try it, and most of them get healthier. "I treat very severe type 2 diabetes," he said, "so the alternative is zero percent get better." In that context, he said, his success rate "is pretty good."

Intermittent fasting as a common tool for weight control has now run far ahead of the research that could establish its safety beyond all reasonable doubt. It's another informed gamble. Like the physicians who now prescribe fasting, the few researchers studying it agree that fasts of up to a day regularly can be beneficial and carry little risk. Many of them do it themselves. (At a June 2018 meeting in Zürich of researchers and physicians studying LCHF/ketogenic eating for type 2 diabetes, hosted by the re-insurance company Swiss Re, I took a poll of the fifty in attendance: More than forty were skipping at least one meal a day.)

If you fast for longer than twenty-four hours, though, the risks of fasting gradually increase, and you have to hope that they don't outweigh the benefits. Jason Fung, who has had as much clinical experience on this issue as anyone, believes longer fasts are effective methods to resolve obesity and type 2 diabetes. Steve Phinney and Jeff Volek of Virta Health, both informed researchers, are less sanguine. They worry specifically about loss of lean mass (muscle rather than fat) when fasting for longer than a day or two or full-day fasting more than once or twice a week. For those with diabetes, longer fasts require that medications be adjusted for times without food and then readjusted when the fast ends. "Improper medication management carries significant health risks," as they put it.

Prescription drug use can also stall weight loss with LCHF/ketogenic eating, and that's another matter that will require the assistance of an informed physician. Certain drugs are known to promote weight gain, and others might. The most obvious are drugs for diabetes—insulin injections, for instance—but if you're avoiding carbohydrates, you'll minimize your need for these drugs. Some anti-anxiety medications and antidepressants can cause weight gain and so inhibit weight loss. Epilepsy drugs can cause weight gain. Some blood pressure medications will do it—in particular, the family known as beta blockers—as will some contraceptives and even antihistamines for allergies.

"You have to look at the benefits and risks of stopping medications or changing them," as Charles Cavo told me. Despite his experience now with fifteen thousand patients, he still called the process of weaning off prescription drugs "a can of worms." For a physician like Cavo, prescribing LCHF/ketogenic eating may also require a discussion with the doctor who originally prescribed the medications and an understanding on that physician's part of the efficacy and philosophy of LCHF/ketogenic eating. All this has to be considered seriously and treated seriously if the goal is to achieve and maintain a healthy body weight and get as healthy as possible.

18

Caution with Children

How should kids eat?

Is LCHF/ketogenic eating suitable for children? Does it work? Is it safe? Like all the issues I've been discussing, little research exists to answer these questions definitively. Again, we have to be guided by common sense. A reasonable anxiety about dietary treatments for children with obesity is that the treatment and the obsessive attention to how and what they eat will result in a permanent or near-permanent eating disorder. The conventional definition of an eating disorder includes "highly restrictive eating," and abstaining from essentially all of an entire food group certainly falls into that category. Most authorities prefer children and adolescents not abstain from carbohydrates for all the reasons they prefer adults not abstain: It's better to remain overweight or obese with a balanced and conventional dietary approach that doesn't work—mild restriction of all calories equally plus exercise—than to be obsessive with a way of eating that might.

Treading carefully is in order. I suggest that children and adolescents who would like to change what they eat to achieve and maintain a healthier weight should do it in a way that is based on human physiology rather than on physics (energy in, energy out) and that has the best chance of achieving their goals.

Since James Sidbury, Jr.'s work at Duke in 1975, it has been clear that LCHF/ketogenic eating works as well for children with obesity as it does for adults. Children can lose weight without hunger and eat to satiety. The academic research literature even includes evidence that LCHF/ketogenic eating induces weight loss without hunger in those with such genetic disorders as Prader Willi syndrome, which is characterized by both extreme fat accumulation and ravenous hunger. ("Food Is a Death Sentence for These Kids" is how a 2015 headline of a **New York Times Magazine** article described the problem.) As early as 1989, William Dietz, then a nutrition researcher at MIT and later the director of nutrition and physical activity at the Centers for Disease Control, was reporting that a low-calorie ketogenic diet was "especially successful" on patients with Prader Willi syndrome who lost significant weight eating it and "whose characteristic ravenous appetites appeared to be suppressed."

But just as adults have to embrace LCHF/ ketogenic eating, to succeed, and maintain it for a lifetime, so very likely will children. For anyone to do that, understanding the rationale is critical. That's a lot to ask of any child, particularly when that reason is controversial and the authorities are arguing that abstinence from carbohydrates does more harm than good. It will also surely help if the parents and the other siblings embrace LCHF/ ketogenic eating as well.

Among the clinicians and other professionals I interviewed, I could find only a handful who specialized in the treatment of children with obesity. Jenny Favret, a registered dietitian, has worked with the Healthy Lifestyles Program at Duke University Medical Center since 2006, when it was founded by Sarah Armstrong, a pediatric obesity specialist. With few exceptions, the Healthy Lifestyles Program admits only children in the top 5 percent of body mass index, who often have weight-related problems (comorbidities) such as diabetes or fatty liver disease as well. Thirteen years later the program had served over thirteen thousand children and families (over 100,000 patient visits), and its staff had expanded to include several pediatricians, physician assistants, nurse practitioners, physical therapists, dietitians, and a behavioral specialist.

For the first five years, Favret told me, the program provided families with conventional dietary counseling: structured meals, controlled portions, low-fat foods, no sugar. A few years in, Favret heard Eric Westman speak, and after her initial skepticism—"What's this guy talking about?"— she gradually started "getting it." She read the available literature and decided the rationale underlying LCHF/ketogenic eating made sense.

By 2011 Favret and Armstrong and their colleagues had created an LCHF/ketogenic diet as a treatment option for the children. As Favret describes it, the eating plan was carefully designed to provide a balance of real foods, heavily focused on low-carb (i.e., green leafy) vegetables, generous sources of protein, and considerable fat via butter, olive oil, coconut oil, heavy cream, full-fat cheeses, nuts, nut butters, and avocados. Fatty protein sources are recommended, rather than lean: cold-water fish, poultry with the skin attached, tofu, and well-marbled beef. (Now, said Favret, she "cringes" at the notion that she used to tell families to eat low-fat foods.) All the obvious carbohydrate-rich foods are initially eliminated, including milk and fruit juices. "To help manage cravings for favorite carb-based foods (and also to minimize dietary boredom)," Favret said, "families are provided with recipes for making delicious alternatives, such as creamy mashed cauliflower, zucchini 'noodles,' cheesy crust pizza, and even various types of fat bombs," the primary ingredient of which is coconut butter.

The ketogenic phase is continued for as long as

desired, at which point some slow carbs such as legumes and whole oats are gradually reintroduced, as well as whole fruit. The focus of the eating plan continues to be large servings of low-carb vegetables, along with adequate protein and a lot of fat. Favret, Armstrong, and their colleagues also teach mindfulness in eating: eating only in response to actual hunger and taking the time to eat leisurely, giving the child time to recognize when he or she is actually satisfied. Whether the families choose to do an LCHF/ketogenic eating plan or to simply focus on eating what Favret described as "a controlled carbohydrate diet of higher fat (real) foods," all are encouraged to enjoy their food and to eat it mindfully.

As Favret explained it to me, many of the families and children in the program see significant weight loss just by eliminating the obvious carb-rich foods and beverages. But those who embrace the full LCHF/ketogenic plan find it works better than any of the alternatives. It does so without advising the children and their families that they have to consciously restrict how much they eat, and without prescribing calorie levels for them. "These kids are just not as hungry," Favret said, "which may be something they've never experienced before. We hear they have more energy. We certainly have many kids whose body mass index is decreasing, which is a success. We have many kids whose liver function tests normalize. That's a success. We have kids whose blood lipid abnormalities improve. They don't just lose weight—they get healthier."

The Duke experience is not unique. David Ludwig, who directed the Optimal Weight for Life clinic at Boston Children's Hospital for twenty years, has had similar success. Of all the patients he and his colleagues have seen at his clinic, Ludwig told me, about a third have little or no interest in changing what they eat. Another third, who take seriously their counseling to avoid sugars, grains, and starchy vegetables, "would lose some weight," Ludwig said, "and then gain a little back. Their risk factors would tend to improve, but it was clearly an ongoing struggle." The last third show "really substantial and sustained improvement, and in those it's really dramatic. When you see these kids a year later, they look like completely different people."

With children as with adults, the key to success seems to be the degree to which they can embrace LCHF/ketogenic eating, remain confident in the approach, and learn to pull the right levers when it's not working. As with conventional family-based therapies for children, success is more likely if everyone in the family, other siblings included, eat the same way and the house is free from temptations. The child who has to watch a sibling eat pasta for dinner and sweets for dessert while she is abstaining is a child who is likely to find abstinence verging on impossible. "If there's a Coca-Cola in the fridge," as Rob Cywes said to me, "a child is going to drink it."

Cywes specializes in bariatric surgery for adults and adolescents, and he often works with children who weigh upward of 250 pounds. He believes that for these children surgery is often necessary to bring their weight under control, but they have to learn to avoid carbohydrates so they can stay lean afterward. When I asked him how he gets his young patients to buy into carbohydrate abstinence, he responded with a question of his own: "How do you eat an elephant?" When I had no answer, he gave it to me: "One piece at a time." He starts with the advice not to drink calories, and particularly sugary drinks, and then moves on to what he calls "vehicle foods": the carb-rich foods we use to transfer other foods from plate to mouth. Instead of sushi with rice, he advises, eat sashimi, without it. Eat burgers without buns. Meatballs but not spaghetti. The inside of a burrito, not the outside. The next step is to give up candy and snack foods. If he can get his patients that far, he makes it a game. "Let's see how deep into ketosis you can get," he tells them. As they start to see and feel the difference, he says, not surprisingly, it gets easier.

The challenge ultimately is the society in which we live. One mother who has a daughter with obesity and asked for anonymity told me their challenge is not just the third-grade teacher who uses cookies or candy to reward performance, or the juice boxes and social pressure at monthly or bi-weekly birthday parties, but her daughter's completely understandable desire to be normal in a world in which eating everything, particularly sweets, is, indeed, the norm. "You even have to choose your words carefully," she told me. "If you say, 'Oh, we do a low-carb diet,' it is suddenly this horrific forbidden thing. If you say, 'Oh, we eat vegetables and meats and healthy fats,' then the response is 'Oh, that's wonderful.'"

Finally, I want to tell you about a young physician and her daughter, who will both remain anonymous. When I asked this physician, to whom I would like to give the last words in this book, what changed her perspective on how to treat her patients, she told me, "The honest answer is that my daughter has obesity. I have been watching her gain weight year after year and seeing her struggle with that and trying to understand it. I have been developing this empathy because she is my daughter. Until then I never had the firsthand experience. But seeing it in my child and trying to wrap my head around what was happening made me think more critically about it." Her own family is slender, she said, but obesity runs on her husband's side. Together they have a son, whom she described as "a beanpole," who can eat anything. Her daughter began gaining thirty pounds a year in fourth grade.

"I didn't know what to do about it," she said, "besides saying 'Don't eat that' or 'Don't have a second doughnut at a school party,' because even though you can control what you cook in the house, even when they're living with you, kids, like everybody, have three million opportunities outside the house to eat crap, especially sugary crap. At that point I brought her to the doctor and had a really bad experience, because the doctors don't know what they're doing either. It's all this 'eat less, exercise more' advice. But nobody wants to say that too much because they're afraid they'll give these young girls an eating disorder, too. It's all very gentle and not very effective."

Carbohydrate restriction and eating "healthy" has helped her daughter maintain weight, but they have yet to restrict carbohydrates sufficiently to see if LCHF/ketogenic eating would really help her daughter lose some of her excess fat. "She wasn't willing to do that," the physician said of her daughter, and she wasn't going to push. Meanwhile, she made the effort to understand the physiology and metabolism and perhaps why her daughter was both putting on fat and always hungry. She found a physician who was willing to help with her daughter. Now she herself has changed how she approaches obesity and type 2 diabetes in her patients. "So much of what I'm doing now is about getting people healthy," she told me, and LCHF/ketogenic eating works with her patients.

"Getting people to lose weight and not be hungry is the key to having any success, and low-carbohydrate, high-fat is the only one that really does that," she said. "People think it's so complicated, but it's not. Such a big part of what I'm doing is trying to get people to be on board, to understand what we're talking about, to stop blaming themselves, stop starving themselves, to follow up correctly and to have the experience of success."

This book is also the end result (so far) of twenty years of research, writing, collaboration, and revolution. As such, any acknowledgments section of reasonable length is bound to be inadequate. The most important individuals contributing to this work, without which none of it would have happened, are the physicians and other medical practitioners who took it upon themselves to try to solve the problem of obesity when the research establishment and the authorities had so conspicuously failed. These individuals have all the attributes that you would want for both doctors and scientists. They are compassionate and curious; they are open-minded and have the courage of their convictions. They did not, as Winston Churchill might have said, stumble over the truth and then pick themselves up and hurry on. Rather, they observed without bias; they generated hypotheses and tested them as best they could. They were less concerned with how they might appear to their colleagues and their peers than with establishing reliable knowledge that might help their patients.

There are now (by my estimate) a few tens of thousands of these physicians worldwide, and their numbers are swelling every day. I am grateful to them all, but I owe a particular debt of gratitude to the earliest adapters, those who assisted my research in its early days when they were only a handful and when speaking to a journalist was likely to do far more harm to their reputation

Acknowledgments

In late October 2016, the journalist Catherine Price joined me for breakfast at a National Association of Science Writers conference in San Antonio, Texas. Catherine has type 1 diabetes, which has given her more than a passing interest, both personal and professional, in how the macronutrients she consumes influence her blood sugar and hence the work of managing her diabetes. She politely insisted at our breakfast that I had to write a book that communicated the messages from my research and my previous work, which informed people both how and what to eat if they are among those on the spectrum from merely fattening easily to fully diabetic and hypertensive. As is invariably the case with Catherine, her arguments were persuasive. This book is a direct result of that breakfast meeting. As I wrote and researched it, it evolved to be something different than we had initially envisioned (for better or worse), but it would not have happened had Catherine not started the process. I am very grateful.

than good. They included Robert Atkins (with all the controversy his name entails), Mary Vernon, David Ludwig, Mary Dan and Michael Eades, Eric Westman, Steve Phinney, and Jeff Volek (a PhD and an RD, not an MD).

For this book, over 140 physicians, dietitians, health coaches, and parents of children with obesity from around the world gave graciously of their time to speak to me about the challenges they and their patients or clients or children face in embracing (or failing to embrace) LCHF/ketogenic eating. I list them here in alphabetical order:

Pedro Aceves-Casillas, Riyad Alghamdi, Richard Amerling, Ahmad Ammous, Martin Andreae, Matt Armstrong, Lisa Bailey, Janethy Balakrishnan, Enrica Basilico, Susan Baumgaertel, Hannah Berry, Ken Berry, Ashvy Bhardwaj, Kathleen Blizzard, Shari Boone, Evelyne Bourdua-Roy, Sean Bourke, Coen Brink, Barbara Buttin, Patrick Carone, Charles Cavo, Aamir Cheema, Kelly Clark, Jonathan Clarke, Zsofia Clemens, Brian Connelly, Kym Connoly, Mark Cucuzzella, William Curtis, Bob Cywes, Joseph Dirr, Carrie Diulus, Susan Dopart, Georgia Ede, Barry Erdman, Vicki Espiritu, Jenny Favret, Sarah Flower, Peter Foley, Gary Foresman, Kyra Fowler, Carolynn Francavilla, Jason Fung, Jeff Gerber, Becky Gomez, Deborah Gordon, Mike Green, James Greenfield, Paul Grewal, Glen

Hagemann, Sarah Hallberg, David Harper, Jennifer Hendrix, Jim Hershey, Birgit Houston, Mark Hyman, Aglaée Jacobs, Rimas Janusonis, Peter Jensen, Bec Johnson, Marques Johnson, Lois Jovanovic, Mirian Kalamian, Katherine Kasha, Fern Katzman, Christy Kesslering, Hafsa Khan, Garry Kim, Kelsey Kozoriz, Janine Kyrillos, Ryan Lee, Dawn Lemanne, Brian Lenzkes, Kjartan Hrafn Loftsson, Andrea Lombardi, Tracey Long, David Ludwig, Unjali Malhotra, Mark McColl, Joanne McCormack, Sean Mckelvey, Nick Miller, Victor Miranda, Jasmine Moghissi, Campbell Murdoch, Daniel Murtagh, Toni Muzzonigro, Ted Naiman, Mark Nelson, Lily Nichols, Brett Nowlan, Robert Oh, Stephanie Oltmann, Sean O'Mara, Randy Pardue, Claire Parkes, Rocky Patel, Charles Pruchno, Lara Pullen, Christina Quinlan, Allen Rader, John Raiss, Sundeep Ram, Deborah Rappaport, Michelle Rappaport, Deb Ravasia, Laura Reardon, Caroline Richardson, Patrick Rohal, Jonathan Rudiger, Amy Rush, Jennifer Rustad, Brian Sabowitz, Andrew Samis, Laura Saslow, Robert Schulman, Cate Shanahan, Ferro Silvio, Michael Snyder, Eric Sodicoff, Sarah Sollars, José Carlos Souto, Alexandra Sowa, Franziska Spritzler, Monica Spurek, Jeff Stanley, Erin Sullivan, Bridget Surtees, Mihaela Telecan, Wendy Thomas, Maria Tulpan, David Unwin, Priyanka Wali, Robert Weatherax, Donna Webb, David Weed, John Wegryn, Eric Westman, Eliana

Witchell, Sue Wolver, Miki Wong, Rick Zabradoski, and Carin Zinn.

I'm exceedingly grateful to those friends, researchers, and physicians who read this manuscript in draft: Mike Eades, Andreas Eenfeldt, Mark Friedman, Sarah Hallberg, Bob Kaplan, David Ludwig, Naomi Norwood, Steve Phinney, Catherine Price, Laura Saslow, Carol Tavris, and Sue Wolver. All provided valuable comments and criticisms, while preventing me in multiple cases from erring badly. The manuscript was made immeasurably better because of their contributions and criticisms. Any remaining errors and failings, of course, are my responsibility alone. With David Ludwig and Mark Friedman, in particular, I have been fortunate to have an ongoing discussion of these issues that never ceases to inform my understanding and challenge my preconceptions.

I would like to thank my extraordinary agent, Kris Dahl at ICM, who has been with me for all my books. I am deeply indebted and forever grateful to Jon Segal at Knopf, who has shepherded all four of my nutrition books into print and given me the confidence that I could say what had to be said (no more, no less). He has become a good friend along the way. I'm also grateful at Knopf to Erin Sellers, Victoria Pearson, Maggie Hinders, Lisa Montebello, and Josefine Kals.

Three institutions have made this work possible

over the years: the Robert Wood Johnson Foundation (for **The Case Against Sugar**), the Laura and John Arnold Foundation (for funding the Nutrition Science Initiative), and CrossFit Health, in particular Greg Glassman, Jeff Cain, and Karen Thompson. I am deeply grateful to all three organizations. I also have to thank my colleagues, present and past, at the Nutrition Science Initiative and especially my fellow board members, Victoria Bjorklund and John Schilling, for their unwavering support, assistance, and friendship.

To my family, Sloane, Nick, and Harry, thank you for everything, with love. Nothing more need be said.

20 the Brooklyn physician: Taller 1961.

20 "a grave insult": White 1962, p. 184.

20 Pennington published his results: Pennington 1954; Pennington 1953; Pennington 1951b; Pennington 1951a; Pennington 1949.

20 a cardiologist in New York City: Donaldson 1962.

21 "Letter on Corpulence": Banting 1864.

21 "more or less rigid abstinence": Brillat-Savarin 1825.

22 the least healthy imaginable: **U.S. News** updates its diet listings online every year, making it difficult to read the versions from previous years online. This reference links to the 2018 press release, which links to the latest available. **U.S. News** 2018.

23 World Health Organization: World Health Organization 2018.

23 U.S. Department of Agriculture: U.S. Department of Health and Human Services and U.S. Department of Agriculture 2015.

23 National Health Service: National Health Service 2019.

23 American Heart Association: American Heart Association 2017.

1 THE BASICS

30 "The Heritage of Corpulence": Astwood 1962.

30 "a brilliant scientist": Greep and Greer 1985.

Notes

The notes below include only the most relevant sources for each chapter. For those who wish to delve further into the background and science of this subject and perhaps deconstruct or challenge the arguments made in this book, please refer to my earlier books—specifically, **Good Calories, Bad Calories** and **The Case Against Sugar**—for (relatively) full accounts of the history and evidence and more detailed annotations.

INTRODUCTION

4 American Heart Association and: Arnett et al. 2019.

5 almost 700 percent: CDC 2014.

6 "conversion": Gladwell 1998.

14 "mass murder": Jean Mayer quoted in Borders 1965, p. 1.

14 "bizarre concepts of nutrition": **JAMA** 1973.

17 "What we see in our clinics": Bourdua-Roy et al. 2017.

32 "To attribute obesity to 'overeating'": Mayer 1954.

33 "lame excuses": Rynearson 1963.

33 "unresolved emotional conflicts": Wilson 1963.

39 "an innate maladaptive behavior": Lown 1999.

40 "amazing how little": Bruch 1973.

42 "the intake of foods": Davidson and Passmore 1963.

42 "Every woman knows": Passmore and Swindells 1963.

2 FAT PEOPLE, LEAN PEOPLE

47 "as if by magic": Gladwell 1998.

47 "the compulsory tendency": Bauer 1941.

48 "What you see is": Kahneman 2011.

50 "not too much": Pollan 2008.

52 "a slow-motion disaster": Chan 2016.

53 "having eaten until fat": Gay 2017.

54 "At one level, obesity": Yeo 2017.

54 "People who have obesity": **Nutrition Action** 2018.

55 "We think regulation of hunger": Kolata 2019.

57 "a perverted appetite": Newburgh and Johnston 1930b.

57 "various human weaknesses": Newburgh and Johnston 1930a.

58 "well known" phenomenon: Stockard 1929.

58 "Probably she": Newburgh 1942.

59 "the most common form": **Time** 1961.

63 "These mice": Mayer 1968.
64 "It's constitutional": Shaw 1910.
64 "I want to be sure": U.S. Senate 1977.
66 "foodlike substances": Pollan 2008.

3 LITTLE THINGS MEAN A LOT

67 "The importance of calories": Groopman 2017.
68 Von Noorden estimated: Von Noorden 1907.
73 "There is no stranger": DuBois 1936.
73 "Why then do we not": Quoted in Rony 1940.
74 "an old trick": Greene 1953.

4 SIDE EFFECTS

80 1,200 to 1,500 calories total: U.S. National Heart, Lung, and Blood Institute n.d.
80 starving a fat man: Sheldon and Stevens 1942.
83 "The best definition": Keys et al. 1950.

5 THE CRITICAL IF

89 "fat and he eats all the time": Goscinny and Sempé 1959.
93 "possibly want to work": Bruch 1957.
96 "without regard to": Wertheimer and Shapiro 1948.

6 TARGETED SOLUTIONS

97 "It is in vain to": Burton 1638.

99 "All diets that result": Nonas and Dolins 2012.

101 "She wakes up, showers": Hobbes 2018.

101 "big benefits": CDC 2018.

102 Diabetes Prevention Program: Diabetes Prevention Program Research Group 2002.

104 "may only make it": Brown 2018.

107 "excessive fatigue, irritability": Ohlson quoted in Cederquist et al. 1952.

107 "literally disappearing": Bruch 1957.

108 his clinical experience: Pennington 1954; Pennington 1953; Pennington 1951b; Pennington 1951a; Pennington 1949.

110 reports similar to Pennington's: Hanssen 1936; Leith 1961; Milch, Walker, and Weiner 1957; Ohlson et al. 1955; Palmgren and Sjövall, 1957; Rilliet 1954.

110 "The absence of complaints": Wilder 1933.

111 Raymond Greene's version: Greene 1951.

111 "Concentrated carbohydrates": Reader et al. 1952.

112 "general rules": Steiner 1950.

113 "gave excellent clinical results": Young 1976.

7 A REVOLUTION UNNOTICED

115 "a revolution in biological": Karolinska Institute 1977.

118 "The fact that insulin": Haist and Best 1966.

119 "just grew fatter": Plath 1971.

120 "the principal regulator": Berson and Yalow 1965.

121 "categorically": Gordon, Goldberg, and Chosy 1963.

122 "I know the math": Gay 2017.

9 FAT VS. OBESITY

133 "High blood glucose": Nelson and Cox 2017, p. 939.

134 Here's how this science looks: Frayne and Evans 2019.

138 "The first principle": Feynman 1974.

138 "nonsense": Borders 1965.

138 "favors fat synthesis": Mayer 1968.

140 "fat mobilizing hormone": Atkins 1972.

140 "bizarre concepts": **JAMA** 1973.

10 THE ESSENCE OF KETO

143 "lobster with butter sauce": Atkins 1972.

146 "to a broad-spectrum": Phinney and Volek 2018.

150 "fat is mobilized": **JAMA** 1973.

150 "the negative stimulus": Berson and Yalow 1965.

151 "it is desirable": Berson and Yalow 1965.

152 "exquisitely sensitive": See, for instance, Cahill et al. 1959.

153 "exquisite sensitivity": Bonadonna et al. 1990.

11 HUNGER AND THE SWITCH

162 "The satiety value": Kinsell 1969, pp. 177–84.

163 "crackers, potato chips": Sidbury and Schwartz 1975.

164 "protein-sparing modified": Palgi et al. 1985.

167 "anorexia": **JAMA** 1973.

168 "Does it help people": Brody 2002.

172 "Before I got": Gay 2017.

173 "even an apple": Pennington 1952.

175 rats can be: Richter 1976.

175 "It is not a paradox": Le Magnen 1984.

178 "star McDougallers": DrMcDougall.com n.d.

179 "We agree people": Krasny 2019.

182 exercise will improve: Holloszy 2005.

12 THE PATH WELL TRAVELED

186 I admitted to trying: Taubes 2002.

190 "unnatural factors": Rose 1981.

192 "vintage fats": Calihan and Hite 2018.

195 "reproducibility crisis": See, for instance, the series of articles published in **Nature** 2018.

196 "Trying to determine": Hecht 1954.

198 "Not everything that causes": Bittman and Katz 2018.

198 "questions also remain": Velasquez-Manoff 2018.

204 three groups of researchers: At Harvard, Taylor et al. 1987. At UCSF, Browner, Westenhouse, and Tice 1991. And at McGill, Grover et al. 1994.

205 "analogous to stewards": Becker 1987.

207 "What we see": Bourdua-Roy et al. 2017.

212 five clinical trials: Brehm et al. 2003; Foster et al. 2003; Samaha et al. 2003; Sondike, Copperman, and Jacobson 2003; and Yancy et al. 2004.

214 "Insulin therapy was": Hallberg et al. 2018.

214 The bottom line was: Athinarayanan et al. 2019.

217 "wishful science": Bacon 1620.

217 one-third less fat: Keys 1952.

218 American Heart Association recommended: Inter-Society Commission for Heart Disease Resources 1970, pp. A55–95.

218 "depth of the science": Koop 1988.

218 "suggestive": Hooper et al. 2015.

219 "less clear": Hooper et al. 2015.

219 American Heart Association: Sacks et al. 2017.

220 "could be used": National Research Council 1982.

220 "convincing": World Cancer Research Fund and American Institute for Cancer Research 1997.

220 "Eat food. Not too much": Pollan 2008.

221 "intellectual phase lock": Alvarez 1987.

225 about metabolic syndrome: Reaven 1988.

226 "obesity and high blood": Lee 2019.

227 "every other diabetic": Joslin 1930.

231 "the major pathogenic": Christlieb, Krolweski, and Warram 1994.

234 "can promote health": Kolata 2018.

236 "how to shop, cook": Moskin et al. 2019.

13 SIMPLICITY AND ITS IMPLICATIONS

242 "redoubtable enemy": Brillat-Savarin 1825.
244 best-selling diet book: Banting 1864.
245 The first one derided: **Lancet** 1864a.
245 "fair trial": **Lancet** 1864b.
248 Nutrition Science Initiative: Schwimmer et al. 2019.
250 supported at Stanford: Gardner et al. 2018.

15 MAKING ADJUSTMENTS

267 "Because protein stimulates": J. Bao et al. 2009.
277 we are restricting the foods: Yudkin 1972.

16 LESSONS TO EAT BY

283 "Eat food. Not too much": Pollan 2008.

17 THE PLAN

306 "You are out of your mind": Donaldson 1962.
317 "a trend beverage": Moskin 2015.
318 "By competing with uric": **JAMA** 1973.
326 "Alcoholics who care about": Taubes 2017.
331 Robert Kemp: Kemp 1972; Kemp 1966; Kemp 1963.
346 "Improper medication management": Phinney and Volek 2017.

18 CAUTION WITH CHILDREN

349 James Sidbury, Jr.'s work: Sidbury and Schwartz
     1975.
349 "Food Is a Death": Tingley 2015.
349 "especially successful": Dietz 1989.

List of References

Alvarez, L. 1987. **Adventures of a Physicist.** New York: Basic Books.

American Heart Association. 2017. "American Heart Association Healthy Diet Guidelines." December 6. https://www.cigna.com/individuals -families/health-wellness/hw/medical-topics/ american-heart-association-healthy-diet -guidelines-ue4637.

Arnett, D. K., et al. 2019. "ACC/AHA Guideline on the Primary Prevention of Cardiovascular Disease: A Report of the American College of Cardiology/ American Heart Association Task Force on Clinical Practice Guidelines." **Circulation** (March 17): CIR0000000000000678.

Astwood, E. B. 1962. "The Heritage of Corpulence." **Endocrinology** 71 (August): 337–41.

Athinarayanan, S. J., et al. 2019. "Long-Term Effects of a Novel Continuous Remote Care Intervention Including Nutritional Ketosis for the Management of Type 2 Diabetes: A 2-Year Non-randomized Clinical Trial." **Frontiers in Endocrinology,** June 5. https://doi.org/10.3389/ fendo.2019.00348.

Atkins, R. 1972. **Dr. Atkins' Diet Revolution: The High Calorie Way to Stay Thin Forever.** New York: David McKay.

Bacon, F. 1620. **Novum Organum,** edited and translated by P. Urbach and J. Gibson. Reprinted Peru, Ill.: Carus, 1994.

Banting, W. 1864. "Letter on Corpulence, Addressed to the Public." 3rd ed. London: Harrison.

Bao, J., et al. 2009. "Food Insulin Index: Physiologic Basis for Predicting Insulin Demand Evoked by Composite Meals." **American Journal of Clinical Nutrition** 90, no. 4 (October): 986–92.

Bauer, J. 1941. "Obesity: Its Pathogenesis, Etiology, and Treatment." **Archives of Internal Medicine** 67, no. 5 (May): 968–94.

Becker, M. H. 1987. "The Cholesterol Saga: Whither Health Promotion?" **Annals of Internal Medicine** 106, no. 4 (April): 623–26.

Berson, S. A., and R. S. Yalow. 1965. "Some Current Controversies in Diabetes Research." **Diabetes** 14, no. 9 (September): 549–72.

Bittman, M., and D. Katz. 2018. "The Last Conversation You'll Ever Need to Have About Eating Right." March. http://www.grubstreet .com/2018/03/ultimate-conversation-on -healthy-eating-and-nutrition.html.

Bonadonna, R. C., et al. 1990. "Dose-dependent Effect of Insulin on Plasma Free Fatty Acid Turnover and Oxidation in Humans." **American Journal of Physiology** 259, no. 5, pt. 1 (November): E736–50.

Borders, W. 1965. "New Diet Decried by Nutritionists;

Dangers Are Seen in Low Carbohydrate Intake." **New York Times,** July 7, 1.

Bourdua-Roy, E., et al. 2017. "Low-Carb, High-Fat Is What We Physicians Eat. You Should, Too." **HuffPost,** October 4. https://www.huffington-post.ca/evelyne-bourdua-roy/low-carb-high-fat-is-what-we-physicians-eat-you-should-too_a_23232610/.

Brehm, B. J., et al. 2003. "A Randomized Trial Comparing a Very Low Carbohydrate Diet and a Calorie-Restricted Low Fat Diet on Body Weight and Cardiovascular Risk Factors in Healthy Women." **Journal of Clinical Endocrinology and Metabolism** 88, no. 4 (April): 1617–23.

Brillat-Savarin, J. A. 1825. **The Physiology of Taste,** trans. M. F. K. Fisher. 1949; San Francisco: North Point Press.

Brody, J. E. 2002. "High-Fat Diet: Count Calories and Think Twice." **New York Times,** September 10.

Brown, J. 2018. "Is Sugar Really Bad for You?" **BBC Future.** September 19. http://www.bbc.com/future/story/20180918-is-sugar-really-bad-for-you.

Browner, W. S., J. Westenhouse, and J. A. Tice. 1991. "What If Americans Ate Less Fat? A Quantitative Estimate of the Effect on Mortality." **JAMA** 265, no. 24 (June 26): 3285–91.

Bruch, H. 1957. **The Importance of Overweight.** New York: W. W. Norton.

———. 1973. **Eating Disorders: Obesity, Anorexia Nervosa, and the Person Within.** New York: Basic Books.

Burton, R. 1638. **The Anatomy of Melancholy.** Reprinted New York: Sheldon, 1862.

Cahill, G. F., Jr., et al., 1959. "Effects of Insulin on Adipose Tissue." **Annals of the New York Academy of Sciences** 82 (September 25): 4303–11.

Calihan, J., and A. Hite. 2018. **Dinner Plans: Easy Vintage Meals.** Pittsburgh: Eat the Butter.

Cederquist, D. C., et al. 1952. "Weight Reduction on Low-Fat and Low-Carbohydrate Diets." **Journal of the American Dietetic Association** 28, no. 2 (February): 113–16.

Centers for Disease Control and Prevention (CDC). 2014. "Long-Term Trends in Diabetes." October. http://www.cdc.gov/diabetes/statistics.

———. 2018. "Losing Weight." https://www.cdc .gov/healthyweight/losing_weight/index.html.

Chan, M. 2016. "Obesity and Diabetes: The Slow-Motion Disaster." Keynote address at the 47th meeting of the National Academy of Medicine. https://www.who.int/dg/speeches/2016/obesity -diabetes-disaster/en/.

Christlieb, A. R., A. S. Krolweski, and J. H. Warram. 1994. "Hypertension." in **Joslin's Diabetes Mellitus,** edited by C. R. Kahn and G. C. Weir. 13th ed. Media, Penn.: Lippincott Williams & Wilkins, 817–35.

Davidson, S., and R. Passmore. 1963. **Human Nutrition and Dietetics.** 2nd ed. Edinburgh: E. & S. Livingstone.

Diabetes Prevention Program Research Group. 2002. "Reduction in the Incidence of Type 2 Diabetes

Our grand business undoubtedly is, not to see what lies dimly at a distance, but to do what lies clearly at hand.

—An aphorism of Thomas Carlyle's embraced by William Osler as the basis of his practical philosophy of medicine

I thought, "Holy moly, this worked!"

—Ashvy Bhardwaj, a British physician, describing her realization that her patient had reversed type 2 diabetes merely by changing what she ate

with Lifestyle Intervention or Metformin." **New England Journal of Medicine** 346, no. 6 (February 7): 393–403.

Dietz, W. H. 1989. "Obesity." **Journal of the American College of Nutrition** 8, supp. 1 (September 2): 13S–21S.

Donaldson, B. F. 1962. **Strong Medicine.** Garden City, N.Y.: Doubleday.

DrMcDougall.com. N.d. "Success Stories: Star McDougallers in Their Own Words." Dr. McDougall's Health and Medical Center. https://www.drmcdougall.com/health/education /health-science/stars/.

DuBois, E. F. **Basal Metabolism in Health and Disease.** 2nd ed. Philadelphia: Lea & Febiger, 1936.

Feynman, R. 1974. "Cargo Cult Science." http:// calteches.library.caltech.edu/51/2/CargoCult .htm.

Foster, G. D., et al. 2003. "A Randomized Trial of a Low-Carbohydrate Diet for Obesity." **New England Journal of Medicine** 348, no. 21 (May 22): 2082–90.

Frayn, K. N., and R. Evans. 2019. **Metabolic Regulation: A Human Perspective.** 4th ed. Oxford: Wiley-Blackwell.

Gardner, C. D., et al. 2018. "Effect of Low-Fat vs Low-Carbohydrate Diet on 12-Month Weight Loss in Overweight Adults and the Association With Genotype Pattern or Insulin Secretion: The DIETFITS Randomized Clinical Trial." **JAMA** 319, no. 7 (February 20): 667–79.

Gay, R. 2017. **Hunger: A Memoir of (My) Body.** New York: HarperCollins.

Gladwell, M. 1998. "The Pima Paradox." **New Yorker,** February 2.

Gordon, E. S., M. Goldberg, and G. J. Chosy. 1963. "A New Concept in the Treatment of Obesity." **JAMA** 186 (October 5): 50–60.

Goscinny, R., and J.-J. Sempé. 1959. **Nicholas,** translated by A. Bell. London: Phaedon Press, 2005.

Greene, R. 1953. "Obesity." **Lancet** 262, no. 6770 (August 1): 253.

Greene, R., ed. 1951. **The Practice of Endocrinology.** Philadelphia: J. B. Lippincott.

Greep, R. O., and M. A. Greer. 1985. **Edwin Bennett Astwood, 1909–1976: A Biographical Memoir.** Washington, D.C.: National Academy of Sciences.

Groopman, J. 2017. "Is Fat Killing You, or Is Sugar?" **New Yorker,** March 27.

Grover, S. A., et al. 1994. "Life Expectancy Following Dietary Modification or Smoking Cessation: Estimating the Benefits of a Prudent Lifestyle." **Archives of Internal Medicine** 154, no. 15 (August 8): 1697–704.

Haist, R. E., and C. H. Best. 1966. "Carbohydrate Metabolism and Insulin." In **The Physiological Basis of Medical Practice,** edited by C. H. Best and N. M. Taylor. 8th ed., 1329–67. Baltimore: Williams & Wilkins.

Hallberg, S. J., et al. 2018. "Effectiveness and Safety of a Novel Care Model for the Management of Type 2 Diabetes at 1 Year: An Open-Label,

Non-Randomized, Controlled Study." **Diabetes Therapy** 9, no. 2 (April): 583–612.

Hanssen, P. 1936. "Treatment of Obesity by a Diet Relatively Poor in Carbohydrates." **Acta Medica Scandinavica** 88:97–106.

Hecht, Ben. 1954. **A Child of the Century.** New York: Simon & Schuster.

Hobbes, M. 2018. "Everything You Know About Obesity Is Wrong." **HuffPost.** September 19, https://highline.huffingtonpost.com/articles/en/everything-you-know-about-obesity-is-wrong/.

Holloszy, J. O. 2005. "Exercise-Induced Increase in Muscle Insulin Sensitivity." **Journal of Applied Physiology** 99, no. 1 (July): 338–43.

Hooper, L., et al. 2015. "Reduction in Saturated Fat Intake for Cardiovascular Disease." **Cochrane Database of Systematic Reviews** no. 6 (June 10): CD011737.

Inter-Society Commission for Heart Disease Resources. 1970. "Prevention of Cardiovascular Disease—Primary Prevention of the Atherosclerotic Diseases." **Circulation** 42, no. 6 (December): A55–95.

JAMA. 1973. "A Critique of Low-Carbohydrate Ketogenic Weight Reduction Regimens: A Review of **Dr. Atkins' Diet Revolution.**" JAMA 224, no. 10 (June 4): 1415–19.

Joslin, E. P. 1930. "Arteriosclerosis in Diabetes." **Annals of Internal Medicine** 4, no. 1 (July): 54–66.

Kahneman, D. 2011. **Thinking, Fast and Slow.** New York: Farrar, Straus and Giroux.

Karolinska Institute. 1977. "The 1977 Nobel Prize in

Physiology or Medicine" (press release). https://
www.nobelprize.org/prizes/medicine/1977/press
-release.

Kemp, R. 1963. "Carbohydrate Addiction." **Practitioner**
190 (March): 358–64.

———. 1966. "Obesity as a Disease." **Practitioner** 196,
no. 173 (March): 404–9.

———. 1972. "The Over-All Picture of Obesity."
Practitioner 209, no. 253 (November): 654–60.

Keys, A. 1952. "Human Atherosclerosis and the Diet."
Circulation 5, no. 1 (January 1952): 115–18.

Keys, A., et al. 1950. **The Biology of Human
Starvation,** 2 vols. Minneapolis: University of
Minnesota Press.

Kinsell, L. W. 1969. "Dietary Composition—Weight
Loss: Calories Do Count." In **Obesity,** edited by
N. L. Wilson, 177–84. Philadelphia: F. A. Davis.

Kolata, G. 2018. "What We Know About Diet and
Weight Loss." **New York Times,** December 10.

———. 2019. "This Genetic Mutation Makes People
Feel Full—All the Time." **New York Times,**
April 18.

Koop, C. E. 1988. "Message from the Surgeon
General." In U.S. Department of Health and
Human Services, **The Surgeon General's Report
on Nutrition and Health.** Washington, D.C.:
U.S. Government Printing Office.

Krasny, M. 2019. "UCSF's Dean Ornish on How to
Reverse Chronic Diseases." https://www.kqed
.org/forum/2010101869165/ucsfs-dean-ornish
-on-how-to-undo-chronic-diseases.

Krebs, H., and R. Schmid. 1981. **Otto Warburg:**

Cell Physiologist, Biochemist, and Eccentric, translated by H. Krebs and A. Martin. Oxford: Clarendon Press.

Lancet. 1864a. "Bantingism." **Lancet** 83, no. 2123 (May 7): 520.

————. 1864b. "Bantingism." **Lancet** 84, no. 2144 (October 1): 387–88.

Langer, E. 2015. "Jules Hirsch, Physician-scientist Who Reframed Obesity, Dies at 88." **Washington Post,** August 2.

Lee, T. 2019. "My Life After a Heart Attack at 38." **New York Times,** January 19.

Leith, W. 1961. "Experiences with the Pennington Diet in the Management of Obesity." **Canadian Medical Association Journal** 84 (June 24): 1411–14.

Le Magnen, J. 1984. "Is Regulation of Body Weight Elucidated?" **Neuroscience and Biobehavioral Reviews** 8, no. 4 (Winter): 515–22.

Lown, Bernard. 1999. **The Lost Art of Healing: Practicing Compassion in Medicine.** New York: Ballantine Books.

Mayer, J. 1954. "Multiple Causative Factors in Obesity." In **Fat Metabolism,** edited by V. A. Najjar, 22–43. Baltimore: Johns Hopkins University Press.

————. 1968. **Overweight: Causes, Cost, and Control.** Englewood Cliffs, N.J.: Prentice-Hall.

Milch, L. J., W. J. Walker, and N. Weiner. 1957. "Differential Effect of Dietary Fat and Weight Reduction on Serum Levels of Beta-Lipoproteins." **Circulation** 15, no. 1 (January): 31–34.

Moskin, J. 2015. "Bones, Broth, Bliss." **New York Times,** January 6.

Moskin, J., et al. 2019. "Your Questions About Food and Climate Change, Answered." **New York Times,** April 30.

National Health Service. 2019. "The Eatwell Guide." https://www.nhs.uk/live-well/eat-well/the-eatwell-guide/.

National Research Council. 1982. **Diet, Nutrition, and Cancer.** Washington, D.C.: National Academy Press.

Nature. 2018. "Challenges in Irreproducible Research." **Nature** special issue. October 18. https://www.nature.com/collections/prbfkwmwvz.

Nelson, D. L., and M. M. Cox. 2017. **Lehninger Principles of Biochemistry.** 7th ed. New York: W. H. Freeman.

Newburgh, L. H. 1942. "Obesity." **Archives of Internal Medicine** 70 (December): 1033–96.

Newburgh, L. H., and M. W. Johnston. 1930a. "Endogenous Obesity—A Misconception." **Annals of Internal Medicine** 8, no. 3 (February): 815–25.

———. 1930b. "The Nature of Obesity." **Journal of Clinical Investigation.** 8, no. 2 (February): 197–213.

Nonas, C. A., and K. R. Dolins. 2012. "Dietary Intervention Approaches to the Treatment of Obesity." In **Textbook of Obesity: Biological, Psychological and Cultural Influences,** edited by S. R. Akabas, S. A. Lederman, and B. J. Moore, 295–309. Oxford: Wiley-Blackwell.

Nutrition Action. 2018. "A Leading Researcher Explains the Obesity Epidemic" (editorial). **Nutrition Action,** August 1. https://www.nutritionaction.com/daily/diet-and-weight-loss/a-leading-researcher-explains-the-obesity-epidemic/.

Ohlson, M. A., et al. 1955. "Weight Control Through Nutritionally Adequate Diets." in **Weight Control: A Collection of Papers Presented at the Weight Control Colloquium,** edited by E. S. Eppright, P. Swanson, and C. A. Iverson, 170–87. Ames: Iowa State College Press.

Palgi, A., et al. 1985. "Multidisciplinary Treatment of Obesity with a Protein-Sparing Modified Fast: Results in 668 Outpatients." **American Journal of Public Health** 75, no. 10 (October 1985): 1190–94.

Palmgren, B., and B. Sjövall. 1957. "Studier Rörande Fetma: IV, Forsook MedPennington-Diet." **Nordisk Medicin** 28, no. iii: 457–58.

Passmore, R., and Y. E. Swindells. 1963. "Observations on the Respiratory Quotients and Weight Gain of Man After Eating Large Quantities of Carbohydrate." **British Journal of Nutrition** 17: 331–39.

Pennington, A. W. 1949. "Obesity in Industry—The Problem and Its Solution." **Industrial Medicine** (June): 259–60.

———. 1951a. "The Use of Fat in a Weight Reducing Diet." **Delaware State Medical Journal** 23, no. 4 (April): 79–86.

———. 1951b. "Caloric Requirements of the

Obese." **Industrial Medicine and Surgery** 20, no. 6 (June): 267–71.

———. 1952. "Obesity." **Medical Times** 80, no. 7 (July): 389–98.

———. 1953. "A Reorientation on Obesity." **New England Journal of Medicine** 248, no. 23 (June 4): 959–64.

———. 1954. "Treatment of Obesity: Developments of the Past 150 Years." **American Journal of Digestive Diseases** 21, no. 3 (March): 65–69.

Phinney, S., and J. Volek. 2017. "To Fast or Not to Fast: What Are the Risks of Fasting?" December 5. https://blog.virtahealth.com/science-of-intermittent-fasting/.

———. 2018. "Ketones and Nutritional Ketosis: Basic Terms and Concepts." https://blog.virtahealth.com/ketone-ketosis-basics/.

Plath, S. 1971. **The Bell Jar.** Reprinted New York: Harper, 1996.

Pollan, M. 2008. **In Defense of Food: An Eater's Manifesto.** New York: Penguin Press.

Reader, G., et al., "Treatment of Obesity." **American Journal of Medicine** 13, no. 4 (1952): 478–86.

Reaven, G. M. 1988. "Banting Lecture 1988: Role of Insulin Resistance in Human Disease." **Diabetes** 37, no. 12 (December): 1595–607.

Richter, C. P. 1976. "Total Self-Regulatory Functions in Animal and Human Beings." In **The Psychobiology of Curt Richter,** edited by E. M. Blass, 194–226. Baltimore: York Press.

Rilliet, B. 1954. "Treatment of Obesity by a

Low-calorie Diet: Hanssen-Boller-Pennington Diet." **Praxis** 43, no. 36 (September 9): 761–63.

Rony, H. R. 1940. **Obesity and Leanness.** Philadelphia: Lea & Febiger.

Rose, G. 1981. "Strategy of Prevention: Lessons from Cardiovascular Disease." **British Medical Journal (Clinical Research and Education)** 282, no. 6279 (June 6): 1847–51.

Rynearson, Edward H. 1963. "Do Glands Affect Weight?" In **Your Weight and How to Control It,** edited by Morris Fishbein, rev. ed., 69–77. Garden City, N.Y.: Doubleday.

Sacks, F. M., et al. 2017. "Dietary Fats and Cardiovascular Disease: A Presidential Advisory from the American Heart Association." **Circulation** 136, no. 3 (July 18): e1–e23, CIR.0000000000000510.

Samaha, F. F., et al. 2003. "A Low-Carbohydrate as Compared with a Low-Fat Diet in Severe Obesity." **New England Journal of Medicine** 348, no. 21 (May 22): 2074–81.

Schwimmer, J. B., et al. 2019. "Effect of a Low Free Sugar Diet vs. Usual Diet on Nonalcoholic Fatty Liver Disease in Adolescent Boys: A Randomized Clinical Trial." **JAMA** 321, no. 3 (January 22): 258–65.

Shaw, G. B. 1910. **Misalliance.** Reprinted Project Gutenberg, 2008. http://www.gutenberg.org/files/943/943-h/943-h.htm.

Sheldon, W. H., and S. S. Stevens. 1942. **The Varieties of Temperament: A Psychology of Constitutional Differences.** New York: Harper & Brothers.

Sidbury, J. B., Jr., and R. P. Schwartz. 1975. "A

Program for Weight Reduction in Children." In **Childhood Obesity,** edited by P. J. Collip, 65–74. Acton, Mass.: Publishing Sciences Group.

Singer, P., and J. Mason. 2006. **The Ethics of What We Eat: Why Our Food Choices Matter.** New York: Rodale Press.

Sondike, S. B., N. Copperman, and M. S. Jacobson. 2003. "Effects of a Low-Carbohydrate Diet on Weight Loss and Cardiovascular Risk Factor in Overweight Adolescents." **Journal of Pediatrics** 142, no. 3 (March): 253–58.

Steiner, M. M. 1950. "The Management of Obesity in Childhood." **Medical Clinics of North America** 34, no. 1 (January): 223–34.

Stockard, C. R. 1929. "Hormones of the Sex Glands— What They Mean for Growth and Development." In **Chemistry in Medicine,** edited by J. Stieglitz, 256–71. New York: Chemical Foundation.

Taller, Herman. 1961. **Calories Don't Count.** New York: Simon & Schuster.

Taubes, G. 2002. "What if It's All Been a Big Fat Lie?" **New York Times,** July 7.

———. 2017. "Are You a Carboholic? Why Cutting Carbs Is So Tough." **New York Times,** July 19.

Taylor, W. C., et al. 1987. "Cholesterol Reduction and Life Expectancy: A Model Incorporating Multiple Risk Factors." **Annals of Internal Medicine** 106, no. 4 (April): 605–14.

Time. 1961. "The Fat of the Land." **Time** 67, no. 3 (January 13): 48–52.

Tingley, K. 2015. " 'Food Is a Death Sentence to These Kids,' " **New York Times Magazine,** January 21.

U.S. Department of Health and Human Services and U.S. Department of Agriculture. 2015. **Dietary Guidelines for Americans 2015–2020.** 8th ed. http://health.gov/dietaryguidelines/2015/guidelines/.

U.S. National Heart, Lung, and Blood Institute. N.d. "Healthy Eating Plan." https://www.nhlbi.nih.gov/health/educational/lose_wt/eat/calories.htm.

U.S. Senate, Select Committee on Nutrition and Human Needs. 1977. **Dietary Goals for the United States—Supplemental Views.** Washington, D.C.: U.S. Government Printing Office.

US News. 2018. "U.S. News Reveals Best Diet Rankings for 2018." **US News & World Report,** January 3. https://www.usnews.com/info/blogs/press-room/articles/2018-01-03/us-news-reveals-best-diets-rankings-for-2018.

Velasquez-Manoff, M. 2018. "Can We Stop Suicides?" **New York Times,** November 30.

Von Noorden, C. 1907. "Obesity." Translated by D. Spence. In **The Pathology of Metabolism,** vol. 3 of **Metabolism and Practical Medicine,** edited by Carl von Noorden and I. W. Hall, 693–715. Chicago: W. T. Keener & Co.

Wertheimer, E., and R. Shapiro. 1948. "The Physiology of Adipose Tissue." **Physiology Reviews** 28 (October): 451–64.

White, P. L. 1962. "Calories Don't Count." **JAMA** 179, no. 10 (March 10): 184.

Wilder, R. M. 1933. "The Treatment of Obesity." **International Clinics** 4: 1–21.

Wilson, G. W. 1963. "Overweight and Underweight: The Psychosomatic Aspects." In **Your Weight and How to Control It,** edited by Morris Fishbein, rev. ed., 113–26. Garden City, N.Y.: Doubleday.

World Cancer Research Fund and American Institute for Cancer Research. 1997. **Food, Nutrition and the Prevention of Cancer: A Global Perspective.** Washington, D.C.: American Institute for Cancer Research.

World Health Organization. 2018. "Healthy Diet." October 23. https://www.who.int/news-room/fact-sheets/detail/healthy-diet.

Yancy, W. S., Jr., et al. 2004. "A Low-Carbohydrate, Ketogenic Diet Versus a Low-Fat Diet to Treat Obesity and Hyperlipidemia: A Randomized, Controlled Trial." **Annals of Internal Medicine** 140, no. 10 (May 18): 769–77.

Yeo, G. S. H. 2017. "Genetics of Obesity: Can an Old Dog Teach Us New Tricks?" **Diabetologia** 60, no. 5 (May): 778–83.

Young, C. M. 1976. "Dietary Treatment of Obesity." In **Obesity in Perspective,** edited by G. Bray, 361–66. Washington, D.C.: U.S. Government Printing Office.

Yudkin, J. 1972. "The Low-Carbohydrate Diet in the Treatment of Obesity." **Postgraduate Medical Journal** 51, no. 5 (May): 151–54.

Index

Page numbers in **bold** refer to illustrations.

A NOTE ABOUT THE AUTHOR

Gary Taubes is cofounder of the nonprofit Nutrition Science Initiative (NuSI). He's an investigative science and health journalist, the author of **The Case Against Sugar, Why We Get Fat,** and **Good Calories, Bad Calories.** He is a former staff writer for **Discover** and correspondent for the journal **Science.** His writing has also appeared in **The New York Times Magazine, The Atlantic,** and **Esquire,** and has been included in numerous "Best of" anthologies, including **A Literary Companion to Science** (1989) and **The Best of the Best American Science Writing** (2010). He has received three Science in Society Journalism Awards from the National Association of Science Writers and is the recipient of a Robert Wood Johnson Foundation Investigator Award in Health Policy Research. He lives in Oakland, California, with his wife, the author Sloane Tanen, and their two sons.